FIVE WARS

For Sergeant Freeman Gardner
and his mother, Farra Ratliff, the bravest among us all.

FIVE WARS
a Soldier's Journey to Peace

FRED JOHNSON

Red Letter Publishing, Austin

Everything herein written is true, but may not be entirely factual.
Some timelines in the book are compressed; some characters are composites of multiple people.
These stories are written as the best representation of the author's personal memories.

Book typeset by Kevin Williamson
Cover design by Kevin Williamson

Created in the United States of America

23 22 21 20 19 18 17   1 2 3 4 5

ISBN 978-0-9981714-9-4 (paperback)

ACKNOWLEDGEMENTS

This book was made possible by the efforts of many people who suffered through the process of teaching me how to best write the stories that are contained in these pages. I doubt that the book would have ever been written had I not met Cathy Fyock, who inspired me to begin the process of assembling my stories and helped me along the way to develop the plan of how to actually publish the book. Cathy also introduced me to Kevin Williamson, who patiently served as my editor and counsel on all things pertaining to the English language.

As anyone who has ever written a book knows, a sentence could not be properly constructed without the service of the teachers in one's life. I could not begin to give justice to all the men and women who inspired my love of reading, writing, and storytelling. However, I could not live with myself if I did not acknowledge the following educators who, in the most fundamental ways, made this book possible: Pat Freeman, Polly Peterson, Edd Fish, Jeff Carling, and John Lane.

The more subtle aspect of completing a book is the counsel of the friends who provide unvarnished and often brutally honest feedback on the quality of the writing. I have Mark Ray, Nana Lampton, Vincent O'Neil, Lois Mateus, Tim Peters—and the most brutal of them all, my wife Dr. Laura Johnson—to thank for that unenviable service.

Finally, there is one author that I must acknowledge. Pat Conroy on died March 4, 2016, during the writing of this book. No other writer had a greater influence on me. Rest in peace, you Prince of Tides.

CONTENTS

He doesn't want to rest

He just wants to run

And he's tired of being told that he's the lucky one.

— *J.G., "War At Home"*

PROLOGUE

I finished my beer and thought: *Where is the best point of departure?*

I was sitting at the bar on the outdoor deck of my favorite hangout in Jeffersonville, Indiana. The 11th anniversary of 9/11 had been the day before, and I had come there to think and reflect on the war, lost friends, and other things. It was warm, and I wore shorts and a T-shirt.

As I looked across the Ohio River to the city of Louisville, Kentucky, nightfall had come and brought with it the brilliant illumination of the downtown skyline. The purple and yellow lights of the Second Street Bridge that connected the two cities merged into a rainbow's array of colors from the office buildings, hotels, and restaurants of Derby City. Louisville was small compared to places like New York or Chicago; however, the urban glow made it appear enormous and bustling. I often got lost in the panorama of the infrastructure's neon peaks, which oddly reminded me of the view I'd had in Kabul of the Hindu Kush and the mountains of Afghanistan.

I had identified two options in Jeffersonville from which I could depart a world that had become too difficult to bear.

One good spot lay a little over a half-mile east of the bar on Riverside Drive, which ran parallel to the Ohio River. It was just past a monument to veterans of the wars in Iraq and Afghanistan. The height of the drop-off to the river was not ideal, and I would have to get enough speed to lift the car beyond a sidewalk and grassy area at the water's edge. I could have gone a little further down the street, where the path to my final destination was more direct and the impact into the Ohio's murky depths was more likely. However, I liked the idea of departing at the monument. After all, I had served in all three wars whose names were engraved in the concrete triangular block: *Desert Storm*, *Operation Enduring Freedom* (Afghanistan), and *Operating Iraqi Freedom*. I had also deployed to Bosnia for *Operation Joint Endeavor*, but that conflict was not recognized on the plaque.

I also liked the western choice. The Falls of the Ohio State Park was about a mile away. A statue of Lewis and Clark marked the entrance to the park and the potential point of departure. It commemorated the explorers' first meeting and the start of their expedition, which would take the Corps of Discovery across the mostly unexplored territories of America to the Pacific Ocean. I greatly enjoyed the book *Undaunted Courage* by Stephen Ambrose, which described the journey, and had found Meriwether Lewis interesting. Lewis was a soldier, frontiersman, and politician. And he was a Renaissance man, something I had hoped to be. He had allegedly committed suicide several years after the expedition, although there remains some controversy over whether he killed himself or he was murdered.

The entrance to the park had the advantage of having a ledge of adequate height and a more direct route into the river. However, the water was shallow there. If it hadn't rained in a while, the limestone rock and coral fossil beds became completely exposed. The impact with the hard surface would mostly likely cause my car to explode and result in an almost certain and fiery death.

There were more effective ways of committing suicide, but I wanted ambiguity, not efficiency. There had to be some reasonable doubt of my intentions. A bullet to the head, hanging, or asphyxiation from vehicle fumes would leave little question of my motive. I liked the idea of going out like Thelma and Louise who, at the end of their movie, drove at high speed off the side of a cliff. The glorious arc and momentary midair suspension would give me one final view of the Louisville skyline. I could be absorbed in it and become a part of its beauty until gravity took me on a downward, headfirst collision with the water, followed—I hoped—by unconsciousness and drowning. A drunken automobile accident would be a plausible explanation of my death. I would rather be remembered as a reckless drunk, not as another sick soldier.

I also chose this method because I didn't want anyone to clean up my mess or have my wife or daughter find me. And I think maybe, subconsciously, I wanted an exit strategy. If, at the last minute, I realized this wasn't such a good idea after all, I could escape as the water poured through the open windows of my car.

I had been developing my plan throughout the afternoon and into the evening while I drank bourbon and beer. The alcohol had filled a great emptiness and stirred up my courage to see the action through, but I did have reservations. I was not quite committed yet.

I decided to do one last reconnaissance of both the east and west options to confirm which would be best. Then—maybe—I would execute my last mission on earth and take my place with my dead friends in Valhalla. I had one last shot of bourbon and paid my bar tab. Then I walked to my vehicle, got inside, and I put the key into the ignition.

MARCH MADNESS

★ ★ ★ ★ ★

It's March Madness 2007 in Baghdad, Iraq.

We come off the Caughman Range on Forward Operating Base Liberty near the Abu Gharib Palace in Baghdad. With the firing stopped we can hear the noontime call to prayer from a nearby mosque. A concrete wall more than 15 feet high separates Liberty from the neighborhood of Monsour, but for me the stirring rhythm of the call to prayer almost make the barriers and space disappear. I was brought to tears once hearing a *muezzin*, the Muslim prayer man, recite the Qur'an in his lyrical cadence before a ceremony; ever since I'll pause to listen when I hear it, even though I don't understand a word.

As the prayer finishes in the background, we're riding an adrenaline rush off the range. We're all oblivious to the 55 pounds of body armor we've been carrying in the rush of an hour's practice at close-quarters marksmanship. As a lieutenant colonel and battalion commander, these are rare moments I can spend with the soldiers where our interactions are as light as the smoke rising from their rifles. It's a cool spring day, just 60 degrees—half the temperature we'll experience in the months to come. The cool air lets the smoke and smell of gunpowder linger on my nose, and I take it in with a deep breath.

Back home, this month marks the real beginning and end of basketball season. It's March Madness and NCAA Tournament time, and the upcoming weeks of games will be our only real entertainment when we're not helping secure Baghdad. As it happens, March will be the height of the Surge in Iraq.

"Can you believe Winthrop beat Notre Dame in the first round?" says Sergeant Dey-Dey Wise, my communications specialist.

I turn and look at Wise through the tinted Oakley protective glasses we wear. With a gloved hand I move the rim of the glasses to the tip of my nose and look over them so he can see my eyes, and I say, "You know I scored 17 against Winthrop my sophomore year."

Dey-Dey shakes his head, tilting it to one side the way he does when he's about to say something he thinks is clever, and quips, "I don't believe it, sir. If you shoot basketball like you do your M4, Winthrop must have been a blind and deaf all-girls school back in the Dark Ages when you played."

I roll my eyes and tell him "whatever," omitting that he wasn't far from being wrong. When I played NAIA basketball at Wofford College in Spartanburg 25 years ago, Winthrop was a few years removed from being a women's teaching school. (As it happens, the young sergeant was also correct about my marksmanship; I am not a great shot and never have been, but I've gotten better over time.)

Even during war, you have to *practice* being a warrior. We spent hours on the range, rehearsing marksmanship drills preparing for our day's work in Baghdad. Not all bases have ranges, so we're fortunate to have one. I can remember when, prior to Desert Storm, while we leaned forward on the border between Saudi Arabia and Iraq, we would just go out and shoot into the great vastness of the desert.

An old Bedouin once stopped our convoy with his herd of goats. He would not let us pass, so we tried to go around him. When he didn't relent, we halted and I got out of my vehicle. The man seemed upset and pointed to a sand dune. I didn't see anything at first, but then I walked over and found a camel keeled over, stiff-legged and dead from our bullets. The old man was on the verge of tears and shook his head like he was asking me why. I couldn't answer him. We never shot blind into the desert again.

On this morning, nine months into my second tour in Iraq, I shoot my pistol and rifle at targets positioned in front of the protected sand embankment that is the Caughman Range. Paper targets are taped to plywood screwed into a 2x4 frame, and we can shoot to our heart's desire. Afterwards, I reload my magazines and do a quick cleaning of my weapons—solitary efforts I enjoy for the time it gives me to reflect.

Warriors must practice their trade because it's not instinctive like it was in the early days of human history, when we had to literally fight to stay alive. There are a blessed few who perform their soldierly duties naturally; Colonel Steve Townsend, my brigade commander, is such a soul who was born for war the way camels are born to endure the desert sands. The desert is, in a sense, in a camel's literal blood; while other mammals have circular red blood cells, a camel's are elliptical instead, which keeps blood flowing even when the animal is dehydrated *and* prevents the cells from rupturing when the camel drinks large amounts of water. Soldiering, likewise, is in some people's blood; Colonel Townsend is one such man.

It is not, however, in mine. I've never been good at shooting guns. Never once have I hunted, and the only time I fired a weapon before joining the Army was when my dad took me to the woods as a youngster to fire a .38 revolver at trees. Even that wasn't for fun; Dad taught me to shoot that pistol, which he kept hidden under the cash register in the bar he owned, to protect our livelihood.

Once, Bill Cooley, who worked the night shift, had to fire it at a robber who thought the bar was an easy target for drug money. Bill hit the thief in the leg, and I asked him if he were just trying to wing the guy. He said drily, "No, I was aiming for his head."

That time in the woods with Dad, I never hit one tree. The pistol's several pounds of metal and plastic felt as alien to me as, well, a camel might feel in Alaska. I recall my hands shaking and the sight moving like the pendulum on a grandfather clock, swinging back and forth across the tree. When I pulled the trigger, my eyes closed and the weapon jerked upwards, almost hitting me my forehead. The bullet tore madly through empty air, snapping tree limbs 10 feet away from the intended target.

If only my marksmanship were as sharp as my jump shot. It's ironic that I'm not a good marksman because I was once a pretty good basketball player, something that, like shooting a weapon well, requires good hand-eye coordination.

Back then, I was all-conference and all-district on the best team in the region. I averaged around 18 points a game—and that was without the 3-point line. If we'd had 3-pointers when I played, I would have averaged well over 25 points, because I rarely shot inside 20 feet. I also made nearly 90 percent of my free throws during games. Once, in practice, I drained 148 charity shots in a row.

I scored over 30 points during several games, launching long-range bombs, turning to trot down court, and leaving a flipped wrist in the air after making the release. I was so sure the ball was going through the net that I was always surprised when I missed. But the last game of my senior year in high school was the worst of the season. My coach, in an attempt to force an inside game against our opponent, would not allow us to shoot from further than 15 feet. We lost to Mount Vernon, one of the worst teams in the conference, in the second round of the regional tournament. I only scored six points—two layups and two free throws. I fouled out in the final minutes of the game, in complete vexation that I suddenly couldn't do what I had been practicing for so long.

As I watch the solders come off the range, I think of that game and the frustration I felt with the constraint my coach placed on me. They take off their body armor, push their kneepads to their ankles, and replace their helmets with boonie caps. They grab bottles of water and find a place to sit, reload their magazines, and joke around with one another. I recognize a similar annoyance with the soldiers that I had felt in that final game of my high school career. Soldiers often complain of the restraint they are placed under by the rules of engagement. This war was different from Desert Storm. Sure, there were rules of engagement then, but the objective of the First Gulf War was clear: get Iraq out of Kuwait. *Kill Iraqi soldiers until Saddam Hussein quits.* It was not hard to identify the enemy in that fight; most of the soldiers signed up for that kind of war.

This generation of soldiers got a counterinsurgency instead; friend and foe couldn't be so easily told apart. Our intention could not be to "kill your way to victory," as General Petraeus, the commander of all troops in Iraq, always reminded them in his speeches. Regardless of what kind of war it is, there will always remain an overwhelming urge to use all available powers of violence against an enemy if that enemy might kills your friends; that's one thing that will never change.

Staff Sergeant Hudgeons, the leader of my personal security detachment, swaggers over and pulls up an ammo crate to sit down. With a well-practiced flick of the wrist, the young Texan snaps a can of Copenhagen snuff, then unscrews the top and, with two fingers and his thumb, pulls out a wad of snuff. With a smooth sweep he plants the fine black tobacco in his bottom lip, dusts off the stray specks from his trouser leg, and puts the can back in his shoulder sleeve. Hudgeons spits and says, "The boys are loaded back up and about ready to go back on the range for quick fire. You good to go, sir?"

I slide the last 5.56mm round into the magazine, slap it against the palm of my hand to seat the bullets, and say, "I'm ready, but I'm getting hungry. What's for lunch?"

Hudgeons smiles and, in his best imitation of Tom Hanks, says, "It's shrimp, Lieutenant Dan."

One of soldiers' running pastimes is finding the right moment to drop a quote from a popular movie. Our most-quoted film, by far, was *Forrest Gump*. Before we depart the base for a mission, Staff Sergeant Hudgeons would always give a brief on the enemy situation, explaining where the insurgents have been fighting and so on. Without fail, his briefs would end with, "And that's all I have to say about that."

Once, while on patrol looking for roadside bombs, we were ambushed by a pair of (inept) insurgents who shot a poorly-aimed rocket propelled grenade at my vehicle. They missed, and the rocket exploded behind me. The attack didn't end well for the enemy, and my gunner Sergeant Knapp said, "Dey-Dey, they shot at you." With his best Gump drawl, and without missing a beat he added, "I'm sorry to break up your Al-Qaeda terrorist party."

Old Gump had it made. He wound up great at everything he did, and he did a lot of things. It wasn't that way for me; even basketball didn't come easy. From the time I was six throughout my teens, I practiced four hours a day, seven days a week, every month of the year. It paid off, and I earned a college scholarship to do the thing I loved most in the world. In high school, I was a shooting guard, but in college I played point. I set the tempo of the game and had to learn to be selfless setting up our better players to score. I became a much better defensive player.

I sought out opportunities to take charges against players a foot taller and 100 pounds heavier than me. I recklessly dove headfirst after loose balls. My knees and elbows stayed red-rashed with floor burns. I would discreetly tap the elbows of shooters as they went up for jump shots to cause the trajectory of the ball to go off course. I would provoke our opponent's best players and throw off their game by trash-talking under my breath.

At the time, I felt no guilt working in this gray area. We were a small school and I personally wasn't as skilled as many opponents, so I sought to level the playing field however I could. Odd though it feels to admit, the experience gave me empathy for the very enemy I curse; they hide roadside bombs and attack us from inside crowds of civilians, packed into the women and children.

I consider the paradox and pull up the black hard plastic pads that now protect my knees. The pads were uncomfortable and awkward the first time I wore them; once, I forgot to put them on before a mission and while on a dismounted patrol, shots were fired nearby and I went to find cover behind a car. Dropping down to a knee, I caught a rock on my kneecap and sprung to stand from the pain, only to be pulled suddenly down to safety by Sergeant Hudgeons.

Tightening up my body armor, preparing to get back on the range, I recollect a game against Georgia Tech. I guarded Mark Price, who would later go on to be an All-American and play in the NBA. When asked about the highlight of my college career, I say, "Holding the great Mark Price to six points." People look at me in disbelief. I then admit: "It was in the second half of the game and he only played five minutes. Price had 24 in the first half, so he still ended up with 30." The joke makes people laugh, but it keeps me reminded of what's true: I was never really a great college basketball player.

Thoughts of going pro were extinguished early in my collegiate career. I started taking ROTC classes and quickly found that my athleticism and the skills honed directing Wofford's offense and defense were skills the Army valued. Going into my junior year of college, part of my basketball scholarship got cut after the coach found two guards who were much, much better than me. I had to find to find a way to make up the difference, so I applied for an ROTC scholarship and got it. I killed two birds with one stone: I got money for school, and I would have a job when I graduated. I only intended to stay in the Army the required time, and then I'd get out. I wanted to be a personnel and administration officer; the Army and my ROTC cadre thought otherwise and I was commissioned an infantry officer. I thought, "The guy who never hunted and hates guns is going to be an infantryman. *Wonderful.*"

I signal to Staff Sergeant Hudgeons to get the soldiers moving. I stand and put on my eye protection and helmet. I walk on the range and begin the second round of our marksmanship drills. I have learned to shoot, and with a lot of practice I've gotten pretty good. Hitting targets is not much of a problem if you take your time; we have optical sights that allow us to put a red dot on a target and the bullet goes almost exactly where the dot is placed. What I don't practice is taking my time. I train to be in a hurried calm, following the shooter's rule of *slow is smooth, smooth is fast.* My rifle is an M4, which as rifles come is small and compact. I can quickly tuck it, aim, and fire. I practice shooting standing up, walking forward, turning, going to a knee, and transitioning from my rifle to my pistol. My M4 is on a sling that allows me to drop it quickly when empty, then unholster my 9mm pistol and fire it. I can change magazines without taking my eyes off the target. If my weapon jams, I can fix it in half a second without even thinking about it.

The day ends with our final firing iteration. Around this time I can hear the third of five calls to prayer from the nearby mosque. We clear the range and climb into our Stryker vehicles to move back to the battalion area; there's something we have to do later in the evening.

This evening, I sit with Command Sergeant Major Julie Walter, the seniormost enlisted member of the battalion. The room is packed. Soldiers stand along the walls and hang back behind the rows of seats. On a stage in front of us is a stand with a M4 turned down on its barrel; a helmet balances on the stock and dog tags dangle from the pistol grip. A pair of boots sit empty in front of the stand. Next to the boots is a picture.

The picture shows a young African-American man. He is sitting with his back rested against an oak tree. His body armor looks almost new; it's pristinely clean. These two things make me think the photo was taken back in Washington state, before we deployed. The soldier's M4 hangs loosely from the strap on his shoulder; tinted glasses (eye protection) sit on his forehead as though he's come from the beach. His lips curl slightly up into a grin—a grin with something more, like a happy secret that sits in his soft brown eyes. The look in his eyes is older than he is. Or rather was, because Sergeant Freeman Gardner is dead.

Five days ago, Sergeant Gardner was pulling security for one of his buddies on a street in the Amiriyia neighborhood about three miles outside the gate of Liberty and Caughman Range. A roadside bomb exploded. He died instantly when fragments of a 122-millimeter mortar passed through his body.

I have been to 18 memorial services since arriving to Iraq, but this is the first time I'm attending a ceremony to honor a fallen soldier from the battalion I command.

We pray, and then some soldiers tell stories about Gardner. I listen to their eloquent little tales of the 26-year-old, newly-married soldier from Little Rock, Arkansas. One of Gardner's best buddies tells a story about the time Freeman gave a professional development class to a group of high-ranking officers, including several generals. The vintage of the officers that attended the class would normally warrant a facilitator of much higher rank than Freeman, who was a specialist at the time. Regardless, Gardner's chain of command determined he was the man for the job.

At the time of the class, Sergeant Gardner's platoon was at a security position on a mountain just east of Mosul. Their location overlooked where the Battle of Gaugamela had taken place in 331 BC, a battle in which Alexander the Great's outnumbered army defeated the Persian forces of Darius III, eventually leading to the fall of Darius' empire. Over the class's hour, the young soldier recounted the dispositions, sizes of the opposing armies, and their strategies and actions on that day. He did this without notecards or a single reference. At the conclusion of the briefing, the generals were given the opportunity to ask questions.

Many of the flag officers felt slighted at having such a low-ranking soldier give a class to them. No one would ask a question, until one general asked: "Where did you learn about this battle, Specialist?"

Gardner pulled out a worn book from his dusty ACU trouser pocket and said, with the greatest respect and sincerity: "Plutarch's *Life of Alexander the Great*. Would you like to borrow it, General?"

The general, normally loud and overbearing, looked perplexed. He sheepishly took the book in his hands and looked into it; he asked, "Does it have any pictures?"

Gardner replied, "I'm sorry, sir, it doesn't, but the words aren't too big."

In the chapel, everyone listening to the story burst out laughing because they knew only Sergeant Gardner had the courage to answer the general so confidently and directly. The soldiers' stories lightened the mood for what was to follow.

In the rear of the room and invisible to us, First Sergeant Cherry, the seniormost sergeant in the company, called out the roll:

"Private First Class Rawlings."

Here, First Sergeant, the soldier responds.

"Sergeant First Class Thompson."

Here, First Sergeant.

"Sergeant Gardner." No response.

"Sergeant Freeman Gardner." No answer.

"Sergeant Freeman L. Gardner, Junior." Silence.

A soldier then stands and says: "He is no longer with us, First Sergeant."

The chaplain reads: "Sergeant Freeman Gardner killed in action on the 22nd of March, 2007 in Baghdad, Iraq."

A pause. Seven rifles fire in unison. Two other volleys follow. While the ring of the shots fades, a trumpeter begins to play taps. As the last note echoes in the night, Command Sergeant Major Walter and I move in front of the boots, helmet, weapon, dog tags, and picture that embody the worldly remains of our comrade. We render a final salute. We salute the *man* that was Sergeant Freeman Gardner, not his equipment. A bagpiper plays "Amazing Grace" and we file out into the night.

I go back to my office, but then head down the hall when I hear a roar from another room. Florida is playing Butler in the Sweet Sixteen, and some soldiers are watching the game, their chairs are as close to the set as possible, their noses almost touching the screen. When the Gators' Taurean Green hits a three-pointer, one of five for the night, soldiers slap high-fives. I become absorbed in the ebb and flow of this fundamental American game. Near the end of the contest, when Green hits another long bomb, my palms begin to sweat and I feel the acid swell deep in my stomach.

I watch the game until the end and then go back to my hooch to sleep. We'll have a counter-IED patrol in the morning, so I'll have to get up well before dawn. Staff Sergeant Hudgeons will give his patrol brief—*And that's all I have to say about that*—and we'll all climb into the Strykers and hit the streets of Baghdad.

March Madness indeed.

BETWEEN THE TRACKS

★ ★ ★ ★ ★

As we drive slowly through the narrow roads leading out of the base, Sergeant Sprahler, my gunner, says over the vehicle intercom, "Sir, I think I've got it figured out how we can finish this war once and for all."

I roll my eyes, though Sprahler can't see because he's sitting below me. I say, "How's that, Sergeant?"

"We tell all the good ol' boys back home that Sadr and Al Qaeda were behind Dale Earnhardt's death," he says. His voice quickens and he says, "So, they'll get in their bass boats with a couple cases of Budweiser and speed across the Atlantic to the Persian Gulf, blasting Toby Keith and tossing beer cans at the aquatic wildlife."

I look down and see Sprahler barely lift himself off his seat and, with closed lips, mimic the drone of an outboard motor, giving the imaginary boat the gas with an extended left hand. Sitting back down, he says, "Then, they'll put up deer stands all over Baghdad and pick off the terrorist bastards one by one. That's my plan. Whaddya think, sir?"

I can imagine him smiling to himself, but I say, "Baghdad's dry. What happens when they run out of beer?"

From the back of the truck, Staff Sergeant Hudgeons squawks into his radio, "They get all jacked up on Mountain Dew, come at 'em like spider monkeys, and scissor-kick 'em in the head!" paraphrasing Ricky Bobby's redneck kids in *Talladega Nights*.

I smile and nod approvingly at Hudgeons's timing. "Good one, Sergeant Hudgeons. Didn't see that one coming." I ask my driver, Specialist Gonzalez, if he knows the location of the battalion we're joining in Monsour. Gonzalez assures me he can get us there as we drive along the shores of Z Lake toward Signal Hill, the two distinctive and pleasing landmarks of our side of the base.

Saddam Hussein created a paradise next to the slums of Monsour just outside what is now the Forward Operating Base or FOB; he built lakes and palaces right next to the Baghdad International Airport. The coalition forces assumed control of the airport—and the landscape's opulent view—upon the invasion. We then built our own concrete walls and fortified it from attack. I've heard that 55,000 U.S. and international military members and civilian contractors reside on the entire complex.

The first day I arrived at Liberty and toured the base, having come from the surrounding areas, I felt like I had arrived in Disney World. To know Monsour, and most other areas of Baghdad, you have to experience them. (And that's all I have to say about that.)

Monsour is the *beladiyah* where we're conducting operations today. Baghdad is composed of a number of major neighborhoods, roughly like New York City; a *beladiyah* is one such area, akin to what we'd call a *borough*, like the Bronx or Brooklyn or Manhattan. There are nine *beladiyahs* in Baghdad, and the city is home to eight million people in total.

Beladiyahs are comprised of big neighborhoods called *hayys*, which are made up of *muhallas*, areas a little larger than city blocks. At first it's all confusing, but once you learn it, having some clear geographic division is very helpful for a military force tasked to clear insurgents and to track the sectarian tug of war that is the battle to secure Baghdad.

I recall my hometown, Centralia, Illinois, and how it, too, was separated into parts, divided by two sets of train tracks. The southern rail line ran to Paducah, Kentucky, less than 100 miles away. South of those tracks was Wamac, home to the industrial park that employed most of the city's population and to 22 of the city's 32 taverns. You could get a beer in Wamac if you were tall enough to see over the bar.

North of town, across a set of railroad tracks that ran all the way to Chicago, was "the Ville." This neighborhood was as black as Wamac was white, and it was home to the best basketball in the city.

My house and the drive-up liquor store my dad ran were in the center of town. I spent my childhood between those two sets of tracks, but whenever the weather was good, I would ride my bicycle three miles north to play basketball at "the Garden." Named after Madison Square Garden, home of the New York Knicks, the Garden featured two full basketball courts on the campus of Lincoln School, one of two nearly all-black grade schools in the Ville.

On late spring afternoons, I stood out on those courts as much for my shirtless white body as for my characteristic inability to jump. When September came around and I was forced back to my neighborhood for the school year, my tan would be deep and dark, even under my arms, and I sometimes believed that I had gotten more spring in my legs.

My dad, who earned eight varsity letters in high school (in the days when only four sports were offered) drilled me in the fundamentals of basketball. He taught me everything he knew—but I learned the most important lessons on the black asphalt of the Garden. Those lessons came slow at first, as I spent most of my time on the sidelines watching, not being picked to play. (I was, after all, the only white kid.) When the games ended late in the afternoon, the exhausted players would retire to sit on the hoods of their cars, gulp brown-bagged quarts of Miller High Life, and stand in judgment of the sideliners who took the court after them.

It was during the later games, I found out, when the older boys took mental note of the up-and-comers and assessed who was talented enough to take the spot of regulars when they weren't there. The prime seat at center court, and the focus of activity for the postgame, was a 1968 sky-blue Mercury Marquis—the throne of Marcus. His stereo blared Parliament, the bass pulsating George Clinton's lyrics to "Give Up the Funk" straight through our bones. Surrounded by his own minor sycophants (and the Ville's most beautiful women), Marcus, the Garden's proverbial Nero, chose who was worthy to take the court with him. A thumbs-down from Marcus sank many a player's career before it ever began.

Marcus looked like African royalty in Illinois: his skin was a deep, dark black and he stood tall at six foot three, but his full afro made him look much taller. On the court, Marcus was more artist than athlete, unorthodox and void of textbook fundamentals. His game was a dance choreographed in real time by his own sense of rhythm, sublime to watch and impossible to imitate. His jump shot release came from somewhere behind his head, without noticeable spin or follow-through—a shooting coach would have pulled out his hair over Marcus's lack of technique—but he couldn't be blocked and he rarely missed. He never played in high school because he considered school optional, but it's doubtful his nature and style would have survived rigorous practice anyway.

One Sunday late in the summer, I was picked to play with the big boys, along with my fellow sideliners Kenny and Odell. We knew that our every move was being watched by the bloodshot eyes of those who'd paid their dues on Sundays long before.

Fortunately for me, there were only ten boys left for the later game. I was matched up against Odell, Marcus' brother. Odell was as skinny as a Somalian, but his elbows were sharp and he knew how to use them. Although Odell was my age, he'd already spent some time in jail; the previous winter he'd gotten a bad case of the munchies late one night, so he broke into a convenience store and stole a case of Twinkies. Shortly thereafter, the police found him passed out at home. Bewildered that he'd been caught committing what he thought an untraceable crime, Odell asked how they had found him, so the police officers took him outside and showed him the footprints in the snow that led straight to his house from the store about a block away. Odell was as mean as he was stupid—and he was not happy that I was on the court.

From his Marquis, Marcus took a long pull from a joint and blew the smoke into the mouth of one of the girls next to him. To start the game, he threw a slick, worn Rawlings basketball to midcourt and simply said, "Ball in." *Let the rite of passage begin.*

Kenny grabbed the ball off its first bounce on the court, checked the ball with his guarding opponent, took two dribbles, and launched a 30-foot jump shot that missed the goal entirely. Kenny looked over at Marcus to see him laughing and slapping high-fives with the other boys. Tyrone, a near equal of Marcus on the court but smart enough to remain on the high school team, yelled "Air ball" several times and continued to taunt Kenny every time he touched the ball. In fairness to Kenny, he wasn't exactly up against champion athletes; with these amateur-hour matchups, the game would become a one-on-one match for whoever just happened to have the ball until one of the teams scored the 12th basket. What I'd learned from watching many of these later games was that *subtlety* was the key to gaining the eye of Marcus. He was the type of critic who saw the strokes in a painting.

I did as my father had taught me and played smart. I moved without the ball, putting myself in good spots where I could retrieve a rebound or open a pass when a selfish player became overwhelmed. This was made difficult by Odell, who used his shiv-like elbows to jab me every chance he got.

Because of the anarchy of these later games, runs would last longer than when the older boys played. They, unlike us, played with a tight sense of purpose; games often lasted only 15 minutes or so. Their reputation, and the affection of the fairer sex, was decided by who won and stayed on the court for the last game. In the schoolyard days, we were a herd, and on this homestead there would be only one alpha male and four studs at the end of the day.

Our games could go on seemingly forever, and we had only one chance to make an impression because the older boys quickly got bored watching our uncoordinated games play out. This worked to my advantage because I was in better shape than the other players, and unlike them I hadn't spent the afternoon drinking beer. As time passed, the volume on Marcus' stereo increased to almost-unbearable loudness, a cue that the boys weren't paying much attention anymore.

It was when Odell finally wore down, winded by chasing me and warped from the reefer, that my opportunity came. The moment was ripe: the score was 11 to 11. *Sudden death.* An errant shot careened off the rim, and I grabbed it near the edge of the court and squared to the basket. Odell, hands on his knees, snorting like a bull, sweat dripping from his tip of his nose, said: "Whatcha gonna do, white boy?"

I pump-faked, drawing him off balance, took a one-dribble crossover, and let loose a 20-foot jump shot that arced through the air and passed straight through the netless rim. For a perfect moment in midair, I had won the game—but before my feet touched earth, Odell's fist drove hard into the side of my face, bringing blood to my tongue and knocking me to the asphalt.

Steadying myself on one knee, one hand on the ground, I used the other to wipe the blood from my mouth. I looked up, my head pounding from both the blow and the bass from Marcus' car. Odell stood defiant and ready, both fists clenched, and he told me to get up. Everything inside me told me not to fight, to stay on the ground, but I stood. Odell closed on me with the smell of booze on his breath, a smell I knew all too well from when the drunks at Dad's place would get too close to my face while asking for credit on their tab.

"Honky, you got no business here. I'm gonna kick your punk ass," Odell growled, drawing back his fist. I stepped forward to close the distance, but before he could throw the punch Marcus' stereo abruptly fell silent, pulling every eye to the Mercury Marquis.

Marcus flicked his cigarette away, slid off the hood of the car, and said as calm as a preacher, "Ain't nobody throwing punches today. White boy schooled you, Odell. You need to stand your Twinkie ass down." A few of Marcus's goons chuckled as the glare on Odell's face curled into a scowl.

Then Marcus continued: "Beaver Cleaver came to the Ville to shoot the rock *with us*, instead of hanging out watching *The Flintstones* back at his crib in Wally World. From here on out, my brothers, Beaver here is my personal guest in the Garden."

Marcus slid back onto the hood and nodded to one of his women, who turned the volume back up. And George Clinton proclaimed to us all, *we love you Dr. Funkenstein, your funk is the best.*

Metallica's *Enter Sandman* is blasting on our vehicle's intercom as we approach Entry Control Point (ECP) 1 next to Caughman Range. I have to scream into the mic of my headset to tell Sergeant Wise to cut off the music. I allow tunes to be played when we are inside the FOB, but not when we are outside the wire. The ECP is a concrete shelter where a lone soldier stands waiting for convoys of trucks to leave. Our group of three trucks slows at the ECP, and Sergeant Strahler leans over the side of the truck; he has a small dry-erase board on which is written the number of vehicles and personnel in the party, then my name as the convoy commander. Strahler came up with the idea to use a dry-erase board because he got tired of yelling over the trucks. The soldier at the ECP writes down the information and gives a thumbs-up, and the gate to Forward Operating Base Liberty slowly begins to open like a medieval castle's drawbridge. We snake through concrete barriers until we're on Route Thunder, which leads to Monsour.

We're no longer in the safety of the base. I lock and load my M4 rifle and 9mm pistol and pass the final checkpoint that separates us from security. Like the flip of a switch, our senses become heightened and the idle chatter on the vehicle's intercom goes silent.

Monsour is one of the last Sunni bastions of Baghdad. Shi'a insurgents from Muqtada al-Sadr's Mahdi Army are pushing from two sides, squeezing the Sunni against the Tigris River to the east. We call Sadr's approach to gaining control of Baghdad Pac-Man tactics: his militia move in, usually under the cover of darkness or at well-placed checkpoints, and convince Sunni citizens it's in their best interest to leave the neighborhood. The Mahdi Army does this one *muhalla* at a time, one after another, gobbling up the city and gaining pre-eminence over the capital as they go. Once the Sunni leave to find sanctuary elsewhere, usually on the rural outskirts, the Shi'a move in to the vacated houses under the protection of the militia. Sometimes there are not enough Shi'a to move in, and the houses remain empty.

However, this is not a one-sided war—far from it—and the Sunni have as much to do with the violence as the Shi'a, if not more. There are more Shi'a living in Iraq than Sunni, though the latter constitute the majority of Muslims throughout the world—and therein lies the rub. Since the creation of Iraq in the 1920s, the Sunni minority have dominated the political process in the country and subordinated the Shi'a majority, along with the Kurdish, Christian, Yezidi, and Jewish minorities. The Sunni insurgents—former regime elements— who don't want to give up their power are one threat to security. The greatest threats, however, are the forces of Al Qaeda, the enemy most post-9/11 soldiers joined the Army to fight. In nearly all soldiers' minds, the point is moot whether Al Qaeda was here before we invaded or not; the fact is they're in Iraq now, and they can't leave. As far as we're concerned, it's better to kill them here than on the streets back home.

We drive down Route Michigan, a road that separates the *hayys* of Khadra and Ameriya. Garbage covers the sidewalks and spills into the street. Sergeant Freeman Gardner died not far from here. Before his death, I had ordered the engineer company to which he was assigned to conduct a sanitation mission to remove the trash so the enemy couldn't hide IEDs. The company commander requested not to do the mission; he wanted to wait until later in the week. He asked for more time to plan since we did not have good intelligence on the area of operation. I held firm and told him to have a platoon clear the street.

My military training had taught me to kill and to prepare myself to die. However, it did not teach me how to deal with death from among the living. It did not teach me how to compose a letter for a grieving parent whose loss no correspondence would correct. After Freeman was killed, I wrote a letter of condolence to his mother. I explained the circumstances of his death, specifically that he died protecting his fellow soldiers. I told her how much Freeman was loved by everyone who knew him. I said I was sorry and that her family was in my constant prayers. But I didn't tell her of my role in his death, nor of the sickening torment I felt because I didn't give his company commander more time to prepare.

I try not to think that I'm the reason Gardner's dead. I push even further from my mind the fact that he was picking up garbage when he died. And I wish I could forget that Freeman Gardner was killed picking up garbage—in Iraq, thousands of miles from home, in a war whose purpose I was beginning to question—but I can't.

That thought is interrupted by gun fire. I see Iraqi police down the street take cover behind their cars. I tell Gonzalez to drive forward so I can get a better angle on where it's coming from. Before I can see it, the firing stops, and Iraqi soldiers start herding a group of 20 or 30 civilians to an abandoned building nearby. We drive to the next intersection, and I get out of the truck with Staff Sergeant Hudgeons and my interpreter First Lieutenant David Abuchallak.

The Iraqis have lined up the civilians and are questioning them. I have Dave ask the Iraqi soldiers what was going on. The Iraqi platoon leader explains that a roadside bomb blew up one of their trucks and that he is certain one of the civilians did it. Through Dave, I ask what makes him think that. The young Iraqi officer says, "They are Shi'a."

We load back up and continue our mission of day, which is to visit a combat outpost in the *hayy* of Ameriya. An important aspect of the plan to secure Baghdad is to occupy areas within the cities so that U.S. soldiers and Iraqis will live in closer proximity to one another, providing more security for the civilians. The idea is to choose the worst place in a *hayy* and literally sleep with the enemy.

Combat Outpost Bannister is surrounded by a 12-foot wall positioned to protect three buildings (abandoned houses) from enemy attacks. The buildings are used for offices for planning, meetings with locals, and controlling daily operations, as well as for living quarters. There is one guarded entrance and a large parking area for vehicles. Two concrete guard towers, positioned on each end of the fortress, overlook the neighborhood. From the towers you can see for several blocks, making it impossible to approach without notice.

I stay the night there and talk to the leaders about their plan to secure the neighborhood with the Iraqis. The insurgents try mortaring us several times during our meeting—a method of attack against which the towers can't defend— but the rounds land harmlessly in the morass of sewage and trash nearby.

The next morning, I drive back to Liberty and find on my desk a letter from Ms. Farra Ratcliff—Freeman Gardner's mother. I open the envelope, and inside is a light blue card with violets, sent in response to the letter I wrote Farra after her son was killed. Standing, I read:

Dear Lieutenant Colonel:

Your letter was received at the right time. I was feeling very sad, and wondering about that terrible day. My daughter brought the mail to me and as I read your letter, my mind was put a little at ease. Thank you for taking the time out of your busy day to console my family. Your letter answered some of my questions. Thank you again and God bless you and all the people you command. I called Sergeant Freeman Gardner my "number one Son." I was very proud of him. He lived to be the best.

Farra Ratliff

I sit down and place the card next to my computer, open, so I can see her handwriting. *He lived to be the best.*

BREAD

The first beer went down fast like a sprint. The next was drawn out like a long walk home, where the time seems not to matter. There was distance between sips of the lager, and that space was filled with reveries of dragons slain on Virgil Mountain.

Joe Fenty, Sergeant Major Dave "Chief" Martel, and I had just finished the Virgil Mountain Madness Trail Run and we were re-hydrating. The race was 18.6 miles of gnarly single-track trail that switch-backed its way through the rocky landscape of Kennedy State Forest near the Finger Lakes in upstate New York. We were stationed about 45 miles farther north at Fort Drum. Virgil Mountain had become a tradition for us as we prepared for the long winter ahead, when snowshoeing and cross-country skiing would replace running as our aerobic outlet.

No one came out of Virgil unscathed. Joe and Dave had bloodied knees, and both my right knee and elbow were skinned up and still bleeding. After the race, we sat on a large log that served as our post-race dining table. As we straddled the fallen timber with our plates of bagels and cream cheese and potato chips, we inhaled the carbohydrates and salt we'd lost during the grueling run. We washed it all down with Yuengling beer, our favorite replenishment fluid. Worn down from nearly three hours on the mountain, we felt life and energy flood back into our bloodstreams.

We were on the edge of a campground that served as the race headquarters. Most people mingled near the food table and scoring tent; others were already moving to their cars to drive home. With soldiers' eyes, we surveyed the terrain and peered deep into the forest. We loved running trails; jumping over rocks and into creeks and scurrying through thickets kept us closer, more keenly aware of terrain's crannies and cracks. We were light infantryman in the historic 10th Mountain Division—this was fun for us.

Joe bit into his bagel. He chewed slowly, then told me, "You ran like you were *possessed* today. I thought for sure you would flame out before the last climb." What Joe didn't know was that I *did* flame out, but I'd made sure I put as much distance between him and me before the 1000-foot ascent in the last miles of the race. I took a long pull on my beer and placed the bottle on the ground between my running shoes. The thick brown mud from the trails was starting to dry and camouflage the blue and gray of my Nikes. Blood trickled down my calf and soaked into my socks.

"Dude, there was no way I was going to let you do to me what happened last year," I said. The year before, Joe destroyed me; he was so strong on the climbs. I was nowhere near him at the end. Joe smiled, tipped his beer in a salute, and said, "Next time."

We ran Virgil on the 21st of August 2001. Less than three weeks later, Al Qaeda would fly airplanes into the World Trade Center and Pentagon. Joe and I were coming out of the post gym and saw the first plane strike the towers on a TV at the front desk. Joe's face went pale.

"I can't believe this," I said.

"My mom works there," Joe said.

We found out later that his mom was all right—but the war was getting personal, and fast.

The war spread from U.S. soil to Afghanistan, then Iraq. It became a global War on Terror and then a global War on Terrorism. Five years passed. Joe and I were both selected to command battalions; we stayed in contact throughout our assignments. Joe returned to Fort Drum to command the 3rd Squadron, 71st Cavalry Regiment.

In early 2006, Joe deployed with his squadron to Afghanistan. I went to Fort Jackson, South Carolina, to command 2-39 Infantry, a basic training battalion. I vowed that I would do my best to train the soldiers that Joe would lead in combat.

Not long into Joe's deployment, my office phone rang. I picked up and immediately recognized the voice, even with the static and crackle of an overseas line. It was Joe calling from Afghanistan. He said he only had a couple minutes to talk, but he wanted to tell me something.

"Guess what, Fred?" my friend said.

"What?"

"Kristen is pregnant. Can you *believe it*?" he told me. I sat back in my chair, put my hands behind my head, and rubbed the sharp stubble of my high-and-tight haircut. It *was* hard to believe; Joe and his wife, Kristen, were both 41 years old. They never mentioned wanting children.

I barely recognized Joe's voice on the other end of the phone; his tone, normally an even pitch (some called him aloof) had a different cadence that day. Joe Fenty— the stoic, hardened warrior—was *giddy*. I imagined him high-stepping around his room, slapping high-fives and pumping his fists as if he'd just scored the winning touchdown of life. Anyone who knew Joe would find the very thought of him showing emotion (let alone doing a *kabuki* dance) strange and hilarious. Even through the phone I could see his smile, red cheeks, and bright eyes. He would tilt his head slightly back when he grinned as if he wanted everyone to see his happiness, that human feeling he held in reserve until it really meant something.

Kristen gave birth to their daughter, Lauren, on April 7. Joe was able to call the hospital shortly after the birth and listen to his baby girl breathe and giggle.

On the 5th of May, Joe was killed with nine other soldiers in a helicopter crash east of Abad, Afghanistan in Kunar Province. Evacuation of the dead was made impossible by the weather. Instead of waiting for the storm to pass, his soldiers chose to climb the mountain to bring Joe and the others down.

I remember getting the email that Joe had been killed. The blood drained from my face and head, rushing away to salve my breaking heart. I read it aloud at barely a whisper. *LTC Joe Fenty leaves behind Kristen, his wife of 19 years, and their 28-day-old daughter, Lauren.*

Joe was not the first friend I had lost in the war, but he was the one I loved and missed the most. A couple days later, I got a phone call from LTC Chris Gibson, a mutual friend. He asked if I was interested in being a pallbearer at Joe's funeral at Arlington. My only question was when I needed to be there.

A number of familiar faces gathered in the chapel at Arlington to carry the casket. There was me; there was Sergeant Major Dave Martel, who ran with Joe and I on Virgil Mountain; there was LTC Eric Kurilla, LTC Gibson, and LTC Chris Cavoli. Chris had escorted Joe's body back from Afghanistan, where they commanded together, and he would give Joe's eulogy.

I did not envy Chris as he approached the stage. *How does one do justice to Joe Fenty with words?* I thought. But over the next several minutes, Chris answered that question with the best, most elegant appraisal of a well-lived life that I have ever heard.

He said, "Where my father's family is from in Italy, there is a saying about bread that speaks not only of the food, but also of a person's nature. If a man is of sterling character, he is said to be *buono come il pane*. It means *good as bread*. Bread nourishes the body and the soul. Bread is life. Bread is good." Chris placed his hands firmly on the podium, his knuckles tense and white and still, and he surveyed the audience for a moment. He paused, then said: "Joe Fenty is bread."

As honorary pallbearers, we followed Joe's flag-draped casket to his final resting place. A team of Old Guard soldiers, whose sole purpose is to bring honor to the fallen, escorted Joe. My friend was carried on a caisson pulled by six horses, majestic animals that looked worthy to pull the carriages of kings. The silence of the walk was broken only by the cadence of the horses' *clip clop* along the road; we walked slowly, almost at half-step, on the immaculate roads of Arlington, floating like ghosts amidst a sea of white headstones. The graves of our country's warriors surrounded us on every side.

When we arrived at the gravesite, the team of Old Guard soldiers retrieved the casket from the caisson and sat it down by Joe's grave. The reverent, almost robotic precision of the Old Guard soldiers was mesmerizing; I could almost become distracted as they secured the flag and stretched it tautly over the casket. Old Glory hovered inches above Joe like a flying carpet that would carry him to heaven.

A chaplain performed the service and gave the benediction. Those who were seated were asked to rise. My head was fixed straight forward towards the casket, but my eyes were on Kristen, who held Lauren in her arms.

Even in that cemetery, a place that honors our nation's noblest heroes, I could imagine no one braver than Kristen Fenty.

An order was given to present arms. Those of us in uniform rendered a salute. A squad of seven soldiers fired their weapons in three successive volleys. A bugle played taps. Two soldiers folded the flag in 13 sharp, precise steps, then handed it to an officer who marched to Kristen. Kristen passed Lauren to a woman next to her.

The officer positioned himself in front of Kristen. He bent slightly, holding the flag with one hand on top and the other below. He held it in front of him, stretched his arms slightly, and passing it to Kristen. In a steady voice, he whispered:

"Ma'am, this flag is presented on behalf of a grateful nation as an expression of appreciation for the honorable and faithful service rendered by your loved one."

The ceremony ended. We gathered in groups, not wanting to leave. I was talking to Chris Cavoli when I noticed Kristen rise from her chair with Lauren in her arms. Alone, she walked over to the gravesite. She stood at the head of Joe's casket. Kristen lowered her daughter until she almost touched the coffin's closed top, and Lauren seemed to move the rest of the way on her own. The baby kissed her father for the first and last time.

I deployed to Iraq two months later.

* * * * *

General Petraeus, commander of all military forces in Iraq at the time, was fond of saying: "You cannot kill your way out of an insurgency." In counterinsurgencies like Iraq, Afghanistan, and Vietnam, the focus is the population, average citizens — and their perception of whether life is better with the insurgents or with the institutional government. If they determine life is better with the insurgents, the counterinsurgents will lose the war even if they win every skirmish.

It's like when a U.S. general said to his North Vietnamese counterpart: *we beat you in every single battle.* The North Vietnamese general replied, "That is true. But it is also irrelevant." North Vietnam prevailed in the conflict because the actions of the United States didn't improve life in the villages and hamlets. In many ways, we made things far, far worse.

It was the day after our attack of Baqubah in July 2007. Baqubah was the seat of government for Diyala Province, just 60 miles north of Baghdad. It was also home to 250,000 people who had endured over a year of brutality from (and nearly impenetrable isolation by) Al Qaeda in Iraq, which had declared Baqubah the capital of the Islamic State of Iraq and implemented strict Sharia law. Our organization, 3-2 Stryker Brigade Combat Team—the Arrowhead Brigade—was sent there to liberate it from Al Qaeda's death grip.

My purpose in Baqubah was unlike the traditional role of soldiers in combat. My role was born of the Brigade Commander's strategy to make our attack of Baqubah relevant—by not only removing Al Qaeda, but also by making life better for the citizens.

Al Qaeda had made the residents of Baqubah miserable with strict enforcement of Sharia law that took away most of the people's freedoms. Still, Al Qaeda was effective at controlling basic needs such as food and water and then denying the people any sense of safety. Al Qaeda's attack on the social and essential-service infrastructure turned the city into a smoking mess (and our artillery and precision rockets during the attack surely didn't help). Electricity was available only sporadically. The water was either inaccessible or contaminated. Sewers had been destroyed by deeply-buried improvised explosive devices (IEDs). Trash had accumulated in mountains on the sides of the roads. The food rationing system had slowed to a complete halt, and that—coupled with local markets being shut down and food imports barely trickling in from around the country—meant malnutrition, hunger, and even starvation were becoming real issues. There were far more people unemployed than working, which made recruitment into the insurgency much easier. Streets were barren: no city transportation was available, and people were afraid to move about because of the lack of security. Combine all of that with an enemy still intent on taking back control of the town, and you become painfully aware that the situation could turn from bad to worse overnight.

Then, there was Baqubah's mayor, Abdullah. Abdullah was a short, well-dressed man who reminded me a lot of Danny DeVito, except he had a full and meticulously-maintained head of hair. I referred to Abdullah as *Sidi*, an Arabic term of respect. He called me *Johnson* because Fred was too difficult. When he was excited, he would say my name several times in a row: *Johnson-Johnson-Johnson!*

Abdullah's office had one desk and one chair, and it was always full of very unhappy citizens. The scene somewhat resembled a Wall Street trading floor just before the closing bell, except here old men with turbans, young men in suits, and even a few women in hijabs screamed their grievances over everything from the lack of electricity to the sewage in the streets.

The crowd engulfed Abdullah. His head would appear and disappear from view, bobbing in the sea of the mob as though he were a buoy in rough waters. I raised my arms halfway up from my sides, palms up. *What in the hell are we going to do?* The mayor emerged from the horde soaked in sweat and pulled me to a side room. Abdullah looked around, put one finger to his mouth to shush me, and said, "Johnson-Johnson," pulling me towards a corner of the room farther away from the crowd.

Then he explained to me, among other things, that he'd been a bus driver before becoming mayor. "I'm the only person that would take the job," he told me. "I'm not sure what we should do."

Perfect, I thought. I can't change the oil in my car, let alone fix the electrical grid of a city, and the guy in charge of that city is a bus driver. We were Humpty and Dumpty with no clue how to put Baqubah back together again.

With the promise to meet again the next day, the crowd dispersed in small groups until everyone was gone.

Abdullah slumped in his chair. I pulled up several crates next to the mayor and sat with him. He signaled to his only assistant to bring chai—simple black tea in this case. After letting the steaming cups cool a couple minutes, Abdullah took a sip and winced. I asked him what was wrong. He said, "No sugar. So bitter."

Staff Sergeant Jared Knapp, the leader of my personal security detachment, was listening. Knapp went out to our vehicle and came back with several packets of sugar. When he came back in, he extended them to Abdullah, and Abdullah looked as if Knapp had just offered him a bar of gold. But Abdullah did not take the sugar; he folded the packets gently back into Knapp's hand and pushed them away. "Thank you, but I can't take these," he said.

I laughed. "Why?" I asked. "It's just a couple packets of sugar."

Abdullah's face turned grim. He said, "We haven't had sugar for months. I know one thing about being mayor, and that is that if I take the sugar, there would be a riot." He told us that Al Qaeda controls the warehouses that have the supplemental food items (like sugar) that they normally get from the government.

Abdullah continued: "Saddam Hussein implemented the food distribution system to counter the effects of the economic sanctions of the First Gulf War." The program of Hussein's he described sounded a lot like the WIC system for low-income families in the U.S. Abdullah sighed. "Iraqis became dependent on the government rations. When Al Qaeda seized the warehouses, they used food as a weapon to control the population. The Trade Ministry in Baghdad stopped transporting food to Baqubah because it was not safe. The warehouses here are nearly empty now."

I asked what other items were part of their usual supplement. In addition to sugar, it included food items like rice, dry milk, and flour. The mayor drew a shaky breath and said, defeated: "If we could only get Baghdad to start shipping the food again. That would keep the people off my back for a while so we can figure out what to do with the water, electricity, and sewage."

A look of epiphany flashed onto Staff Sergeant Knapp's face. He took the sugar packets back out of his pockets, one in each hand. He brought the two together into one hand, and as he did he said, "Well, why don't *we* go down to Baghdad, get the food, and bring it back here to Baqubah?"

Abdullah and I had been staring blindly into the abyss. Knapp's suggestion snapped our heads upright. Abdullah and I locked eyes, then smiled and nodded at each other.

★ ★ ★ ★ ★

Carl Von Clausewitz once said, "War is very simple, but the simplest things in war are very difficult." He was right a hundred times over in Iraq, a thousand times over in Baqubah.

Abdullah arranged a convoy of 20 civilian Iraqi trucks to pick up Baqubah's most-needed ration staple, flour, from the Ministry of Trade in Baghdad. He figured that, if we could show the government that it was safe to bring flour into the city, the other food items would be able to follow, and the public distribution system would slowly be re-established. But we had to get that first step right—it was most important to show the city's residents some sign of immediate success, that the situation was getting better and not worse.

Abdullah would not go at this task alone. We would provide security, along with the Iraqi police, to ensure the civilian trucks got to the Ministry of Trade safely and then returned to Baqubah with the flour.

In Baqubah, while preparing the convoy to Baghdad with the Iraqi Army and civilian truck drivers, I was directing traffic trying to get vehicles in position. As I narrowly avoided being run over by a Toyota pickup truck, I saw a young Iraqi man walking briskly towards us, an intense look in his eyes. I saw a bulge in his loose, un-tucked shirt.

Suicide belt, I thought.

I unholstered my 9mm and pointed it at him. I shouted in Arabic for him to stop and lie down, but he kept moving. I shouted again, this time in English: "Damn it, get down!"

He stopped, raised his hands, and in very good English replied: "It's okay, I'm a policeman! I'm here to help!" I told him to raise his shirt. He did, exposing a pistol—which I then found out had no bullets in it.

Akmed, as he introduced himself, said he knew how to get to the food distribution point in Baghdad. Akmed said he would lead us there and help negotiate the release of the food. He left to get a vehicle and some of his friends and promised to be back soon.

The episode continued as the convoy of 20 civilian pickup trucks, an Iraqi Army platoon, a U.S. Stryker platoon, and the three vehicles from my personal security detachment all prepared to move to Baghdad. We waited for Akmed, but he didn't show, so we finally decided to take off.

Baghdad was only 60 miles to the south, but it might as well have been 500. Every 10 miles there were checkpoints, which slowed our progress.

The food distribution center was on the northeast side of Baghdad on the edge of Sadr City. We had plotted the location on our GPS, which should have taken us directly to the location. However, when we arrived at the location on the GPS, there was only a simple market. We kept driving until we'd done three laps all the way around Sadr City. We were about to start our fourth lap when Staff Sergeant Knapp, who was providing rear security, radioed me that a car was approaching fast and flashing its lights. Again, our first thought was that it was a suicide vehicle approaching for attack, but Knapp noticed that someone was hanging out the window of the car waving and pointing for us to turn right. It was Akmed, and yet again he only narrowly avoided getting shot.

We stopped and the young Iraqi pulled up next to us. I asked where he'd been. Akmed said he had to borrow a vehicle and get gas; we had left by the time he'd returned. With little time to waste (because the distribution center would be closing soon), I told him to lead the way. He pointed to a large building a block down the street and told me, "It's right there. Just pull in."

We spent the next several hours trying to negotiate release of the food with the center's manager. The fat bureaucrat wouldn't budge. He said that we had not brought the correct paperwork and we would have to come back in another week.

We didn't have a week to wait. I wasn't leaving and I made that fact clear. The real reason we were denied food was that Baghdad's food distribution center was run by Shi'a Muslims, while Baqubah was predominantly a Sunni town. It was sectarianism, plain and simple. I felt the bile rising in my stomach and my face getting red. I lost my temper. I began to yell, which was a terrible mistake; in Arab culture, men are particularly offended when someone raises their voice to them.

The fat man's security closed in. About the same time, we heard an explosion outside the building. Someone had thrown a grenade at one of our Strykers. The team leader in the vehicle that was hit said everyone was all right, but he told me that we should move soon. A crowd was starting to form outside the gate of the parking lot. We temporarily stopped the meeting so we could assess the situation while Akmed stayed behind.

When we returned, Akmed stopped me in the door. He said, "Everything is under control. I talked to the manager and I know some of the same people. He will release the food, but you cannot lose your temper again. Please."

True to his word, Akmed had worked it out. I learned later that Akmed was Shi'a. I'm sure a deal of some kind was made, but I didn't care.

The next day the food was loaded on the trucks. We returned to Baqubah with the Iraqi Army in the lead as conquering heroes. We had Akmed to thank for it all—the same man we nearly shot twice.

★ ★ ★ ★ ★

Abdullah's stock as mayor grew after our successful mission to secure the food. He went with us on the mission, and he was in the front vehicle upon our return, leading the way into the city. With newfound confidence and the support of the city's residents, Abdullah started making progress fixing many of the other problems in the city.

We'd intended to work our way up Maslow's famous hierarchy of needs— survival and security first, comfort and opportunity later. However, given the devastation that Baqubah had endured, the best we could hope to do was satisfy the basic requirements for living, along with some security.

But Abdullah had more on his mind. He wanted for Baqubah to be self-sufficient, to not rely upon Baghdad for flour. He wanted to re-open the flour mills that had been closed during Al Qaeda's occupation. The mayor told me that the mills had sustained hundreds of jobs. With the mills open, he said, people could work and Baqubah could feed itself. "Johnson," he said, "this is the most important thing we must do."

It may have been the most important, but it was also the most difficult. The largest mill was isolated on the southern tip of the city and would require security forces to defend. It hadn't operated in over a year and it would need to be repaired first. We had to find the people who could fix, maintain, and run the machinery. Not least of all, we had to get wheat to that mill to make the flour.

Everything would work against us except the will of Mayor Abdullah.

We were in Baqubah nearly two months. On the final day before heading back home, Abdullah rode with us to the mill. We entered the building to the hum of machines processing wheat into flour. The chief engineer approached us, smiling, and pointed to containers overflowing with the processed grain. Then, from behind his back, with both hands, he presented us loaves of freshly-baked bread made from the flour produced that day. He handed a loaf to me; I broke it and gave half to Abdullah, who was quietly crying. Abdullah held the loaf to the sky, blessing it.

He looked at me with jubilation. "Johnson-Johnson, this is bread. This is good."

"Yes it is, Sidi. It is good."

COMING HOME

We began our final descent and started to break through the clouds. I leaned out of my seat, craning my head toward the window around a young man slouched against the wall, sleeping. He looked terribly uncomfortable, but he was smiling in his sleep, and I knew why. I was careful not to wake him, but I just *had* to see around him. The dense white began to fade like a large cotton ball coming apart in strands, and I saw the first patch of green. Then, I was hit in the face with a big shamrock glove and knocked giddy with the emerald brilliance of the mountains of Washington state. I stayed there as long as I could, bent over, twisted around the napping passenger, and took it all in until my legs cramped (from 20 hours on the flight) and I had to sit back down. I closed my eyes and imprinted the vision in my mind. If I shut them even now, I would see those same beautiful green mountains again.

There was very little green in the place from which the plane had taken off, except near the great rivers where I'd spent the last 15 months. Those lands had their own unique beauty and magnificence, flush with a history both ancient and mysterious. The place was sometimes harsh and horrific. America went there five years ago in search of nuclear weapons, but the only weapon of mass destruction we found was the most destructive of all: blind, ignorant human hatred. We did our best to dismantle the death machine that had rent the country apart, and I felt good about what we had done. My time was well spent, but there had always been something missing.

The absent part . . .

When I went overseas, it seemed that my five normal senses had gone to sleep; I was left only with my soldier's sixth sense that kept me alert and signaled the signs of danger or the opportunities to turn situations around. At the first sign of green through the plane windows, all five awakened from their long hibernation. When they woke up, it wasn't the gradual process of a lazy Saturday morning; it was waking up Christmas morning as a kid, knowing I had been good and that Santa's bounty would be waiting under the tree.

Seeing the mountains and green on the flight in, I made a plan for reintegration back into a life unfamiliar. It was a simple plan, but a start: I would run those green peaks every chance I got. That was the extent of the thought—no more details except a map of all the parks in Washington state. I got up early on my first full day stateside and drove to the nearest park, anything with rolling green hills.

My drive took me through Tacoma and east of Seattle to a place called the Issaquah Alps. I parked my car, stowed my trail map, and started running up Cougar Mountain. The climb burned my thighs, as I'd been accustomed to flat land, but I still had strength from months of carrying body armor, and I continued up the mountain at a slow but steady pace. I took some risk running then because my eyes didn't spend time looking at the trail, watching for secure footing as a smart runner's would; instead, my eyes were drawn to every tree and moss-covered rock rising out of the earthen soil. I breathed the fresh smell of the forest through my nose and mouth and, it seemed, every part of me.

I considered that everything making this wonderful scent was once alive but had long been dead and decayed; it's ironic such beauty comes from death, that a living being would be energized by the smell.

I thought about Joe Fenty. We used to run similar terrain together in upstate New York. He died on a mountain in Afghanistan. I prayed that some good would come of his loss and put his memory away for another day.

I passed a brook with a small waterfall, and as I listened I could understand why that sound is mimicked and sold on CDs by the thousands; it's sublime, inimitable. The climb continued, but I refused to slow down and walk, though it probably would have been easier. I kept my head and eyes up now, searching for the summit where I'd find reprieve and finally regain my oxygen.

When I finally reached the top, fog obscured the view, so I walked the crest with my hands behind my head, breathing until my chest slowed down and relaxed. Nearby, I saw a box nailed to a post; there was a notebook in it, and travelers would sign the notebook and put it back. I opened the box, read a few messages others had left, and wrote in my own shaky handwriting, "Thank you, God."

I descended the mountain, retracing my steps, until I found my way to the parking lot and into the warmth of my car. On the drive back, I stopped at the No Bull Tavern for a beer. It was still early and there was only one other customer, a man I learned lived on Cougar Mountain, on land his grandfather had homesteaded before Washington became a state. His eyes were a spectacular shade of green; he had snow-capped hair and wise, leathery skin with ridgelines and deep valleys, somewhat like the landscape in his backyard. I ordered a stout, watched the toffee-colored head form during the pour, and listened to the old gentleman talk about beekeeping and salmon fishing, about the coming winter on his mountain and his children, grandchildren, one great-grandchild. Satisfied that I'd heard all he had to say, he asked me: "What did you miss most when you were away in the war?"

I took the stout in my hand and drank long and slow, treasuring the maple, barley, and smoke. Then I sat the glass down and studied him for a long while before I finally said, "You."

BACK ACROSS THE RUBICON

Eyes barely open, I peeped at the racecourse through an opening I'd made in my sleeping bag. I'd climbed into the sleeping bag in my car and started the heater, flooding the cabin with glorious warmth. My eyes fluttered, barely open.

A couple hours before, around 5 p.m., I'd come to my car to warm myself. I'd called my good friend Gary Griffin, one of my running mentors who'd first put this idea—racing on foot for 24 straight hours—into my head. I told him I was still in the race but was taking a break. I hoped I sounded convincing on the other end of the line, but on my end I was still shivering hard and I had no intention of walking another step. I watched the shadows of competitors pass, most of them with their heads hung, staring just beyond their toes and marching through the cold Washington night. In the fog that came before sleep, I cursed myself for not finishing the task I'd started.

I came to Longview, Washington, on the 15th of March 2008 to complete the Pacific Rim One Day Run. The goal of the race: stay in motion for one entire day, from 9 a.m. to 9 a.m. It was a goal I'd set for myself since returning from Iraq six months before. The historical significance of this particular day did not occur to me until the start of the run; just before the race director counted down and said *go*, a fellow contestant warned the other runners toeing the line: "Beware the Ides of March!"

In 44 BC, a fortuneteller warned Julius Caesar to be cautious of this very day. The Roman emperor didn't listen, and he died by the hands of his best friend, Brutus, and his other close associates. Prior to his assassination he was offered the crown three times by Marc Anthony and three times he declined. The Ides of March was forevermore a day that would symbolize impending doom. The personal irony of this didn't come to me until much later in the race, when I was death-marching the course along the shores of the Lake Sacajawea and mumbling *Et tu, Gary* in reference to my close friend Gary Griffin who first put the idea of running 100 miles in my mind several years ago.

I had returned from Iraq after a 15-month deployment, but I was forced to remain separate from my family for another nine months while my wife, Laura, finished her post-doctorate work in Tallahassee, Florida. During that time, I lived in a one-room apartment on Fort Lewis, Washington. My quarters were conveniently located less than a mile from both the gym and my office. Fort Lewis offered great running routes; in less than an hour I could find myself on the sublime trails around Seattle's mountains. I had a lot of time on my hands those days, and I chose to spend it in places like those until I could be reunited with my wife and daughter.

I read that Teddy Roosevelt would attempt something profoundly strenuous after experiencing a setback. After losing to Woodrow Wilson as the "Bull Moose" candidate in the 1912 presidential election, he took a boating trip down the Amazon. The malaria he caught nearly killed him, but according to him, the time on the river regenerated his spirit.

I could account for my own loss in Iraq by repeating the name of Sergeant Freeman Gardner. Gardner had been killed in Baghdad one week after the Ides of March, the year before the Pac Rim. I could keep it that simple if I wanted to. My encounters in combat had changed me in ways I didn't yet see; for the time being, I chose to run.

My training plan for the Pac Rim was to complete a series of races up to six hours and, when possible, double up marathons and ultras on the weekends. Several Saturdays and Sundays, I collected 60 to 70 miles. When I couldn't find a race to run, I would wake up at 5 a.m. and drive the hour to my favorite place in the mountains and run the trails for three or four hours. Inevitably, I would end up at the No Bull Tavern after the workout for Guinness and their delicious venison chili. Sometimes, after eating, I would do another (shorter) run to get accustomed to moving with food in my stomach, as I would have to during the Pac Rim. The beer offered a unique twist to my training regime; it set me slightly off-balance, which (I told myself) I was sure to experience after 12-plus hours of movement.

However, after one training run, I did not have the happy ending of a post workout beer. I went to the Olympia State Forest to run some particularly difficult trails and reach several points that had spectacular views of other mountain ranges in the area. My intention was to finish up at one of the many microbreweries in the city of Olympia.

I set out from a base camp with a trail map, several quarts of water, some energy bars, and my cell phone. I figured I would run for four hours. Somewhere along the way, I misplaced the trail map and I got lost. The four-hour run turned into an eight-hour death march. I ran out of water and ate all my energy bars by the time it got dark. I had no cell service. I started walking, listening for streams to find water. I considered making a lean-to out of limbs and vegetation and wait until morning when I could see.

I had a tinge of fear—the healthy, animal kind that heightens your senses and quiets your conscious mind. It was the same kind of feeling I would get entering a bad neighborhood in Iraq—the kind that makes your eyes dart left and right, scanning for anything out of the ordinary and makes your hearing sensitive to any sounds of warning.

It was then, after eight hours in the forest, that I heard the sound of a car not far away. I sprinted, with what energy I had left, towards the noise of the car's engine. When I emerged from the forest, I could see that the driver had stopped off to the side, in the dark, and was digging through his glove compartment for something. I knocked on the windshield, which startled him; he rolled the window down slightly and the smell of reefer seeped out. I told him I was lost and needed a ride back to my car, that I'd give him $20 if he helped me out. "Sure, dude," he said.

As we drove to my car at the bottom of the mountain, an odd feeling of sadness washed over me, like I'd missed something by not staying the night in the woods. I'd missed that fear.

A race like the Pac Rim would strike fear into many hearts, but not mine. I had my strategy for the race, and it was simple: I would pace at eight-and-a-half-minute miles for as long as I could. When I couldn't sustain that pace anymore, I'd slow to nine-minute miles. I would slow the pace this way continually until eventually I was walking, and then I would do that as long as I could. Of course I'd eat and drink, but most importantly, I would stay in motion for 24 hours.

Most runners on the soft gravel track wore iPods, but daydreams were my entertainment. Most of the time, I focused on my form and pace. I created rewards for myself—like taking a 30-second walk break at the aid station—but mostly I just ran with an empty mind, save the single mantra of "reserve your energy" that I repeated to myself. I mostly ran alone for the first six hours on the course; I talked to others to be hospitable, but I made sure I wasn't dragged into their pace, nor into their own wallowing in pain and dread.

As I circled the loop those first hours, I recalled a poem my father had me memorize as a kid growing up in Illinois.

> *It's easy to run at the start of the race*
> *With a free and easy tread,*
> *To keep in the pace and follow the pack*
> *While the whole track lies ahead.*
> *But when the finish line heaves in sight*
> *And the veins and the muscles swell*
> *And the eyes are glazed—*
> *That's the time when class will tell.*

Where I'm from in the Midwest, there's not much to do. There's corn, basketball, and a lot of taverns (my dad ran one of the 32 bars in our town of less than 10,000). For two consecutive summers, a local charity group staged a bicycle marathon in the parking lot of a mall; they figured it would attract a lot of kids and bring some amusement to the hot July day. The rules were straightforward: ride your bike as long as you can without stopping or touching the ground. The first year, when I was 11, I won by riding for 18 hours.

The second year brought tough competition. It was down to me, Kevin Kramer, and Mike Schulte, grade-school classmates of mine.

Dad came out at 6:30 a.m. before work to see how I was doing. We'd been riding for almost 20 hours and I was about done. My father told me to recite the poem he taught me. So I did—I spoke it from memory as I peddled, barely keeping my bike upright. I repeated the lines as loud as I could so he would hear me across the track. Kevin and Mike stuck close to me as I shouted the words at the top of my lungs.

I passed my dad again and he said, "I get off at 4:30 and I'll see you then. Mom will be by with some food at lunchtime." He clenched both his fists and raised them in front of himself like a boxing champion. He said, "Be strong, son. When the going gets tough, the tough get going." Kevin and Mike looked at each other, took one more lap, and then stopped. I think they knew I wouldn't disappoint him—so they knew they wouldn't win.

Just shy of the Pac Rim 50-mile mark, the thought occurred to me that I was entering new territory; I had never run more than 50 miles at once. I tried not to think much about it. Still, I couldn't help but ponder the explorers from the old world, who imaged that the edge of the earth—or monsters at least—might lie just beyond the horizon. I'd gotten farther than I'd ever been, but I knew that each passing mile that evening would bring me closer to Hell.

About then, I met Dave.

Dave was 55 and a computer consultant, with multiple Western States 100-mile races under his belt, including one in 19 and a half hours some years back. He came from behind me during one of my walk breaks, took a few steps with me, and simply asked me, "Are you ready to run?"

I nodded—and so began four hours of an impenetrable calm, moving at a pace much faster than I had managed the previous two hours by myself. I felt as though Dave and I were the lead horses of a stagecoach, hitched side by side. Time passed in waves. I had no clue we sped up, only that I felt right about it.

Night came and street lights illuminated the course. I ate bananas and boiled potatoes with salt until I couldn't eat any more. I switched to pizza and then turkey sandwiches. I drank water, Gatorade, and my favorite: good ol' Mountain Dew. Dave and I would walk a minute or so while we ate and then we'd resume running. I thought to myself that I had only been hungrier once in my life, and that was in Army Ranger School. I told Dave about it.

I'd left Fort Benning on the 26th of May 1986 looking like Skeletor. My face would have been perfect for a high school biology class studying the human skull. I was all cheekbones, eye sockets, and chin; my skin stretched over my body like plastic wrap. Every vein (and what muscle I had left) was fully exposed. I think that was the last time I ever had a six pack.

I lost 35 pounds in a little over two months of Ranger School, and when I left, I was determined to gain back every pound as fast as I could.

The day after graduation from Ranger School, I drove from Fort Benning, Georgia, to Wofford College in Spartanburg, South Carolina. I had been invited to administer the Oath of Office to the newly-commissioned second lieutenants of the same ROTC program from which I'd graduated a year before.

The 300-mile trip—which usually took five hours or so—took me eight hours this time. I stopped at every McDonald's, Krispy Kreme, and Dairy Queen along the highway. I steered with my elbows almost the entire way, using both hands to shove ice cream and fried bread into my mouth. By the time I arrived in Spartanburg, my stomach had partially taken the shape of a volleyball and jutted out from my black and gold "Rangers Lead the Way" T-shirt like a pregnant woman's belly.

I met Tommy, a college buddy, at Clancey's Pub and asked him to lead the way to the bar for a pitcher of Bass ale. Tommy was a year behind me at Wofford and we had been in ROTC together. He would be in the commissioning ceremony the next day and then he would take off for the Infantry Officer Basic Course. Tommy planned on going to Ranger School. He was very interested in my thoughts on the Ranger training experience. Over beer (and numerous servings of tater tots) I told him how I survived on one meal a day, how I did a thousand flutter kicks in training, how I marched mile after mile with a heavy rucksack.

To get through tough times, I'd write things on the inside of my patrol cap (which we called a "PC"), quotes that would help me feel better when I read it. The inscriptions were made at great peril. If the Ranger instructors discovered that I denigrated the sacred PC, I would be forced to wear an empty sandbag formed into the shape of a sailor's hat on my head. But at night, under my poncho, with a red-lens flashlight, I carefully scribed phrases that I'd remembered from favorite books or some of the influential people in my life. When I got feeling a bit fragile, I would remove my patrol cap and review my inscriptions one at a time.

From *Jonathon Livingston Seagull,* I pondered that "The perfect speed is being there."

Camus gave me: "One life is as good as another; all you have to do is change your habits."

When things got especially tough, my old man's words motivated me just as they had when I was a kid. He used to say, "When the going gets tough, the tough get going."

I always found myself returning to Thoreau, however, and his reflections on the year he spent living on Walden Pond. The quote I chose was long and wrapped all the way around the bill of my PC. Thoreau said—and I wrote: "I went to the woods because I wished to live deliberately, to front only the essential facts of life and see if I could not learn what it had to teach."

At first, that night over beers, I had Tommy's rapt attention as I recounted the hailstorms, the waist-deep swamps, the night rappels, and the airborne jumps. However, as I prattled on, embellishing my own little epiphanies, I could see his eyes glaze over.

Then, I noticed a girl standing next to our table. She'd been a year behind me at Wofford, and I'd tried unsuccessfully to court her most of my senior year. She was far friendlier than I had remembered, and I invited her to sit down with us. The three of us shared a pitcher of beer, and as the evening progressed she asked if I would like to go with her to another bar. Tommy was nodding off, and it was getting late. I considered it for a moment. I said that I was sorry, but I had another obligation.

As I put Tommy in a taxi, the girl and I exchanged numbers. Inside the taxi, the driver asked, "Where to?"

I watched the young lady walk back towards Clancey's. Before opening the door, she turned, waved, and frowned flirtatiously, her bottom lip stuck out in exaggeration. *Sad face.* I smiled at her and waved back as she walked into the bar.

I told the cabbie, "Waffle House, please."

I would never see that girl again. In the dark on that mile-long track in Longview, Washington I tell Dave, "In Maslow's hierarchy of needs, food trumps sex every time."

Dave laughs and we slowed to a walk for a bit. I told him about the other real challenge of Ranger training, which was the cold. There were times then that I thought I'd never feel warmth again.

It was about that time when the temperature around the track dropped. I got cold—*real* cold, real fast. I was only wearing a soaked cotton T-shirt. I told Dave that I was getting "Ranger School cold." He agreed and said he was stopping to change clothes. I said I would do the same. At this point, I'd run 70 miles.

I veered off the course towards my car as Dave continued along the track to his drop bag. I wouldn't see him again until after the race.

At my car, I immediately turned over the ignition and blasted the heat. I changed my shirt, gloves, and cap in the car to get warm. I put on a GoreTex shell and made a couple phone calls to Laura and Gary, letting them know I was still alive. I shoved jelly beans and M&Ms into my mouth and got back to the track.

I can't say exactly how long I was in my car, but when I started running again I returned to an ominous quiet. Before there had been people and now there seemed to be none, apart from the volunteers at the aid station. Lap after lap, I didn't see a single runner. I began to walk. The street lights didn't seem as bright anymore, and I struggled to spot the landmarks ahead of me. My head down, I walked alone, accompanied only by my thoughts.

Those thoughts turned dark. I thought about the Ranger Class of 1995 and the four students who didn't make it out alive. It had been unseasonably cold that year during the school's Florida phase. The students had been walking for hours in the Yellow River, which had risen considerably from recent heavy rains, and they could never get dry. Four stopped to get warm and broke contact with their patrol. They were found dead of hypothermia the next morning.

There is something to the saying that fatigue makes cowards of us all. The soldiers that died then weren't cowards. You could say, though, that the extreme conditions caused them to make poor choices. The instructors responsible made errors of their own judgment that put the students at risk. But while those four succumbed to the elements, the rest of the class survived.

One of the reasons people, particularly soldiers, take up ultramarathon running and other fanatical physical activities is to practice courage. In any threatening situation, quitting can't be an option. I've always wondered: what is it that separates survivors from those who perish under the same conditions?

I thought of *Into the Wild*, a book by Jon Krakauer. It's about a 23-year-old kid who, after graduating from Emory University, took off in his car and drove west with the intention of hiking alone into the Alaskan Yukon. He made it into Alaska, but he never made it out. He crossed the sort of line where, in the end, "risking it all" really does mean risking it all.

Caesar did the same when he led his army across the Rubicon River, into Rome proper, in 49 BC. Caesar's was an act of treason punishable by death under Roman law—but he accomplished his military objectives, avoided reprisal, and was eventually declared dictator of Rome anyway. Had he failed, he would have been publicly executed—but that didn't happen (at least not right away).

I wondered if I had crossed my Rubicon, some point of no return, by running too fast at the start of this race. Had I used up too much of my strength in the early stages of the run and left nothing for later? I had no proverbial legion to protect me. Doubt began to close in from the edges of my mind. I wanted to quit, which, to most people, would have made sense after 70 grueling miles. But with this sort of challenge—and with the sort of person who accepts it—quitting isn't an option. To release yourself from your battle *after* crossing that Rubicon of the mind, to let go and walk away from a race when there's race left in you—to relinquish the very prize you came there to claim—is the same as dying. And yet, in that moment, I wanted to quit.

But then, Leslie came along. She startled me because I had not seen another runner in such a long time.

Leslie was running her 52nd marathon or ultramarathon of the year. She had raced a tough 50K trail race that morning and wanted to attempt what extreme distance runners call a "double." Leslie was running, barely, but was intent on finishing another 31-mile race. If nothing else, her story of endurance shamed me to keep moving and not feel sorry for myself. We didn't talk much. The "run" crept along at a 12-minute-mile pace, just a bit faster than the walk I'd had going before.

There's not much more I can say about Leslie. We suffered in silence together, but her presence extinguished my doubt. She finished the distance she'd set out to finish as I started my 97th lap.

I knew I was going to make 100 miles, but I would do it walking. When I reached the century mark at 19 hours and 47 minutes, I lied to the lap counter, telling her I was going to take a break and would be back in a little bit. I put aside my goal of moving for 24 hours straight and instead settled for the lesser standard of running 100 miles in a day. That's when I went back to my car, pulled out my sleeping bag, and called Gary. I fell asleep shortly after hanging up the phone.

I can't remember if it was the sun rising at dawn or the thought of asphyxiating on the car's fumes that woke me, but something dragged me out of deep sleep. I got out of my car and stood and stretched. As I did, I thought about going to the aid tent to get some coffee and then coming clean with the lap counter that I'd finished. Standing beside my car, with my arms folded across my chest, I looked out on the course in the blue morning cool.

I saw an old man with whom I'd visited during the run; he was treading his way up an incline that, to me, had become a mountain in my last hours on the loop. I waved at him and asked how he was doing. He didn't look up, but he said loud enough for me to hear:

"This is a timed event, son. Twenty-four hours. It ain't over until 9 a.m. By my watch, it's seven o'clock. You still got some runnin' to do."

The crusty old fart.

I laced up my shoes and ran the bugger down. I kept going for an hour and a half without stopping, even to drink. I ran eight more miles, finishing the last one around a nine-minute pace. I concluded my 108th mile just shy of the 24-hour mark.

I looked for the old man and found him sitting on a cooler drinking a beer. I asked him how many miles he'd finished. "A hundred and five," he said before taking a long pull on his brew. The gentleman looked remarkably like Ernest Hemingway, only leaner; salty sweat filled the deep wrinkles on his face and a blue Seattle Mariners baseball cap sat cockeyed atop his head of bushy gray hair. As I sat, I thought of what he'd said a couple hours before, and he seemed to read my mind.

"Works every time," he said. "There's always something more. We just have to be reminded of that now and then. But maybe I should've kept my mouth shut another hour. Then you'd never have caught me."

He raised the beer and said, "Good run, young man." I thanked him and turned to leave Longview.

In the car, on my drive back to Fort Lewis, I listened idly to the radio and replayed the race in my mind. Though I was very proud of what I'd accomplished, I felt that I'd cheated myself out of the full experience of a 24-hour race. My intention had been to stay in motion for the full day. I'd wanted, if I'd been able, to recreate the deepest fatigue I'd ever felt: Army Ranger training, when the harrowing physical trials and extreme sleep deprivation were enough to conjure hallucinations of imaginary green demons. I once saw a fellow Ranger student try to put a quarter in a tree; when I asked him what he was doing, he replied, "The damn soda machine is broke." I hoped I'd be *that* tired again.

I didn't get there, to that point of hallucination. I didn't achieve the goal I'd set— to stay in motion for an entire day and experience the strange, hollow comfort of complete exhaustion.

Three times during my journey around Lake Sacajawea, I began to cross the Rubicon of my mind, to feel that my only paths led to death or glory. Three times, total strangers brought me safely back to shore.

GOING

My daughter Madelyn shook me out of a deep sleep. "Daddy, Daddy, wake up! Wake up, Daddy!"

I sat up, still in a fog. "What's wrong, honey?"

She spoke so fast I could barely understand her. "He's dead. He's dead! Bin Laden is dead! They killed him!"

I swung my feet to the side of the bed and stood up. She took my hand and dragged me to the television playing the news in the living room. We watched as the broadcaster repeated that Osama Bin Laden had been killed in Pakistan during a raid conducted by the United States.

The significance of this began to soak into my mind as I reflected on our decade-long war—a war that started with a massacre masterminded by Bin Laden. It was the reason I'd spent those 15 months in Iraq three years ago.

Before long, my wife joined us in the living room and sat down next to us. "I can't believe it," she said, "but what does this mean?"

Madelyn looked at her mother as though she'd lost her mind. "Well, it means Daddy doesn't have to go to Afghanistan. Right, Dad?"

My 13-year-old daughter hugged me tight at my neck, her cheek pressed by mine. She was close enough to feel the warmth of her breath. I just put my arm around her and stared at the TV. I curled my lips upward so it looked like I was smiling, but I didn't say anything.

After a minute, Madelyn slowly released me. She shifted on the sofa so she could look me in the eyes. I'd always told her that the eyes don't lie.

I sensed her exhilaration shifting to something else. We looked at one another. Her enormous brown eyes began to swim in tears.

Try as I might, mine did too.

Persistent, she tugged at my shirt sleeve and said, "But the war is over, Dad. We killed him. That means you don't have to deploy." The tears rounded below her eyes, then fell and ran down her cheeks. I still didn't say anything.

"You shouldn't have to go," she sobbed, breaking into tears. I kissed her cheek, and as I did I could taste the salt. When I spoke, all I could say was: "I'm sorry, darling."

Three months later, my wife pulled the car to the curb at the Louisville airport, Madelyn in the back seat. I stepped out, took my bags from the trunk, and put them on the curb. "It's like pulling off a Band-Aid," I'd told them. I was ready then; I didn't cry. They didn't either.

I turned back from the curb, and they were waiting. We circled into a group hug. As I held them, I kissed them both and whispered, "I love you both so much."

As we released, I said to them, "Before you know it, I'll be back for R&R, and we'll be at Derby." I waved goodbye as they got in the car and drove away. Then, I picked up my bags and walked into the airport.

A man saw my uniform and approached me as I waited to check in. He wore a black cap adorned with rows of ribbons colored silver, purple, blue, and red above the bill and the words "Vietnam Veteran" embroidered in gold along the top. "Just getting back?" he asked me.

In the intervening three years, I'd seen men and women like me, in uniforms like mine, with bags like mine, in airports all across the country. I always asked the same thing he did. Now, I was one of the travelers in uniform again.

I shook the old warrior's hand and smiled. "Going," I said.

MON DANABASHEE

My brain hurts—literally. I don't have a headache. It's something much different and far worse. I feel like someone took my brain out of my skull, gave it nerve endings, and chained it to a treadmill running at high speed to rub raw all morning.

The meeting has gone on for three and a half hours, and my mind is exhausted. We've been in the Chief's conference room all morning discussing the Afghan National Army's personnel accountability process. The system itself is complex, yes, but it is made all the more difficult because it is being explained through an interpreter. That might have been fine, but the interpreter isn't doing a good job; the bungled words and meaning have to be made clear by General Karimi, who is chairing the meeting along with my boss, Canadian Major General Michael Day.

General Sher Mohammad Karimi is Afghanistan's Chief of the General Staff. He is the seniormost officer in the Afghan National Army, and for the next year he will represent the very reason I am in the country.

My official position is "Senior Advisor to the Afghan National Army Chief of the General Staff." It is a lofty, and utterly ridiculous, title. Who am *I* to advise the most senior military leader of a sovereign nation, a man who has served his nation for over 40 years? I'm a conduit for information and sometimes a sounding board; hopefully, if I'm successful in building our relationship, I will be a trusted confidant one day. But at the moment—realistically—I am Karimi's Coalition assistant. Nothing more.

Still, as far as I'm concerned, I have the most important job in the war.

My job is to facilitate and enhance the exchange of both information and intent between General Karimi, his staff, and my headquarters at the NATO Training Mission in Afghanistan. In my 26 years of service, I've never had a more challenging job, including plenty of assignments where people were trying to shoot me. More than anything, however, it is the stakes that weigh on my mind.

We have been at war here for nearly ten years. We have invested a great deal of national treasure in Afghanistan, a country nicknamed "the graveyard of empires" because world powers from the ancient Greeks to the Soviet Union have failed to conquer it. With the Coalition drawdown beginning, this next year will be crucial. If we cannot implement a viable plan, we will become another tombstone in the imperial graveyard.

Our side knows the importance of the coming year, and so do the Afghan leaders. Part of the reason for meeting today is to get an *absolute* accounting of the number of soldiers in the Afghan army. Afghanistan's success relies on the creation of a capable, affordable, and sustainable army. Every soldier counts—and the count has to be right.

Throughout the conference, General Karimi has stopped the interpreter to correct his translation. He'll explain what is happening to his staff in Dari, then he'll clarify what is being said to us in English (which is, by the way, impeccable). Karimi is impressive, but this is a tiring process, to put it mildly. Still, we can't cut corners; General Karimi believes poor translation is the reason for most of the problems our armies have working together. "Communication and understanding is everything," he likes to say.

General Karimi is a 65-year-old father of five and grandfather of three. He studied in the United States, Great Britain, and India, and he is qualified by the U.S. Airborne, Ranger, and Special Forces. During the Soviet occupation of Afghanistan, he spent considerable time in a Russian prison because of his affiliations with the West.

General Karimi's first name, Sher, means *lion*. The derivative of his last name, Karim, means *kind*. He is, indeed, a gentle lion.

In the ultimate display of kindness, he suggests that we take a break and resume the meeting at another time. Just like that, my brain pulls the big red STOP plug and the treadmill comes to a halt.

It's Ramadan, which means the day ends early for the Afghans and my work is done at the Ministry of Defense. I walk General Karimi to his office and say goodbye to his personal staff. Then, I walk down the stairs of the headquarters building. At the entrance, an honor guard stands at attention, in his dress uniform, weapon at his side; in accordance with protocol, I salute him and walk out of the building.

"*Khudaa Hafiz*," I say to the guards outside.

One guard—my favorite—says, "Goodbye, Johnson. *Mon DaNabashee*."

I have heard the phrase before. It's a traditional Dari expression, but I'm not sure what it meant. I stop and try to repeat it back: "moo dan nashee."

"No, no, no. *Mon da na ba shee*," he corrects me, repeating the words very slowly. This is our daily drill, part of my education, but my brain is just too tired.

"Let's work on it tomorrow," I tell him.

He shrugs. "Okay. Tomorrow."

Back at Camp Eggers, the U.S. compound where I live, the next part of my day begins. I have several more meetings there. Then, I have reports to write. I work well past my normal bedtime, and, when I finally get to the barracks, my bunkmate is fast asleep, snoring like a thousand pigs swirling in a tornado. Eventually I fall asleep.

The next morning I can barely get out of the rack. My morning run is slow and painful. My 15-minute walk to the Ministry takes me 25. I'm dragging, but I try to shake it off. *Come on, Fred. Gotta be sharp today.*

I have a list of things to discuss with General Karimi. We are helping build an Officer's Military Academy in Afghanistan, similar to Great Britain's Sandhurst or our own West Point, and some matters important to construction of the building require decisions from Karimi. There is also the new Mobile Strike Force Vehicle we have to discuss. And we need to talk about these issues specifically because he has to prepare for his meetings with the seniormost Coalition leadership later in the day.

I arrive at the entrance of the headquarters. My much-adored guard is waiting for me. "Sub Ba-khair," I say with the morning's greeting.

"Good morning, Johnson. *Mon DaNabashee*," he says in return.

This time, I catch the syntax and pronunciation. "Mon DaNabashee," I repeated back.

He smiles, "*Bali*. Yes, Johnson." The other guards clap and laugh.

So I say it again, confidently: "Mon DaNabashee! Mon DaNabashee!"

It occurs to me that I have no idea what I'm saying. For all I know I'm calling myself names; I wouldn't be the first to be tricked this way. I ask my friend, "What does it mean?"

He takes my hand in both of his. He shakes my hand slowly and looks at me with gentle, empathic eyes. Finally he says: "May you not be tired."

MAYBE

I walk out the gate of Camp Eggers and bid good morning to the guard. *Sub Bakhair*, I say, employing one of the few phrases of Dari I've learned since arriving to Afghanistan nearly a week ago. He smiles at my fumbled attempt to speak his language and politely thanks me by replying *tashakur*. Colonel Tom McDonald and I begin our trek through Kabul's Green Zone to the Ministry of Defense. Tom is the advisor to Abdul Rahim Wardak, the Minister of Defense and General Karimi's boss. Tom and I are officemates. A fellow infantry officer, like myself, Tom is an immediately likeable person. Though we have only just met, I hope to be his friend. For now, we are "Battle Buddies." You cannot walk anywhere in Afghanistan alone. It's a good policy, but one that will later cause challenges in my work. Tom and I are going to a meeting where General John R. Allen, the Commander of the International Security Assistance Force (ISAF), is chairing a "Senior Security Shura" with his staff and Afghanistan's defense leadership.

The walk to the Ministry is less than a mile, and the 15 minutes it usually takes pass quickly as I take in the scenery. We walk on the sidewalk along a wall that separates us from President Karzai's palace. Far in the distance, beyond the confines of the Green Zone, I see houses built into the side of brown mountains, the stony structures stair-stepped, one seemingly atop another, until they reach the summit. There doesn't seem to be a road or even a trail, and I wonder how the residents get to the homes at the top, much less how they get water there.

I recall my first deployment, which was to Honduras when I was a lieutenant in the 10th Mountain Division nearly 25 years before. The mountains in Honduras were like the mountains of Afghanistan, except that in Honduras the hills are green and lush with vegetation. Hondurans often passed us carrying pails of water as we trekked up and down the goat trails on patrol. I assume water is transported a similar way here.

At the entrance to the Ministry of Defense complex, we clear our weapons and sign in. Once past the gate and within the grounds of the compound, there are gardeners tending to the flowers, and I'm struck how green and beautiful it is inside. I have several more opportunities to practice my Dari with other pedestrians walking along the way. It's practice, I figure, and by the time I leave in a year, I hope to be able to correctly pronounce most of the phrases I learn.

I've made a commitment to learn the language and culture of the Afghan people during my time here. I regret not doing so in my last deployment to Iraq. Taking in the language and culture makes for a much richer experience. In Bosnia, for example, the maids who cleaned my room made me coffee, and every morning we'd spend time talking. By the end of my time in the Balkans I was decent at Serbo-Croatian, although I've forgotten most it now. Language is a skill you have to practice. Culture, on the other hand, is something that stays with you. Most values transcend cultures; being a good person is all that matters in the end.

We arrive at the door of the headquarters building and are met by a potpourri of international soldiers from Canada, Great Britain, Australia, and a handful of other countries; they belong to personal security detachments for generals of their respective nations. They are loitering, just killing time until the end of the meeting. We greet them and move through the tall glass doors. Once inside, we salute the honor guard and walk a couple flights of stairs to the main conference room. The short climb leaves me out of breath; I'm not yet accustomed to Kabul's elevation (6,000 feet above sea level).

Inside the conference room, I pick up a transmitter with headphones. The device helps bridge the language barrier; as someone speaks, a translator converts Dari to English and speaks through the headphones, allowing those of us who cannot speak the foreign tongue to understand what's being said. I put the earphones on, dial into the proper channel, and wait.

The room is full except for two seats at the head of the table. Colonels—like me and a couple others—are the lowest-ranking officers in the room. The crowd quiets in anticipation of the entrance of the meeting's co-chairs. The door opens, and we stand as General Allen and Minister Wardak enter the room. Wardak speaks perfect English, so I don't need the headphones while he and General Allen exchange introductions.

The day before, a CH47 was shot down, killing 31 American and seven Afghan Soldiers. Both men comment, and General Allen starts by saying, "We not only shed our blood, we *share* our blood." I ponder those words for the rest of the day; I doubt now that I'll ever forget them.

The meeting's course moves from topic to topic, but it remains focused on one central theme: the formation of an Afghan Army that is capable, affordable, and sustainable.

Somewhere near the end of the meeting, I get a sense that I have a front-row seat to the creation of a nation, that I am witnessing something that is, overwhelmingly, much larger than myself.

After the meeting, Tom and I walk a different route back to Camp Eggers. The route we take puts us on a busy street where young children are selling bracelets and scarves. Tom has been in Afghanistan a little longer than me; he is well-known to the kids and guards. That doesn't surprise me considering his engaging personality and kindness. They all greet him along the way, and I see he is much better in Dari than I am. The kids try to peddle their wares to him, but he shakes them off saying, *Nay, tashakur.*

Near the gate, before entering the camp, I meet a little boy who says to me in decent English, "Buy a bracelet."

I tell him, "Maybe tomorrow."

Then the little boy points to the name tag on my uniform and says, "Hey, new guy, what's your name?"

"Johnson," I tell him.

"Okay, Johnson, I remember you," he says before turning to walk away.

I turn to walk through the gate, but then call after the boy to ask how he knew I was new to Kabul. He turned back to answer me.

"Because you say maybe."

KAREEM

Deployment math is tricky.

The most significant calculation a soldier makes—and arguably the only one that matters—is the amount of time he is into a year-long tour of duty. When you first get to country, you round up the number of months you've been there. For example: if I've been in Afghanistan 47 days, that's almost seven weeks, but if someone asks how long I've been here I'll tell them "two months."

Somewhere near the midway point of the tour, though, you start to round down. If I have 47 days *left* in Afghanistan, that's not quite seven weeks; I'm liable to say I have "a little over a month" left. That's what happened in all three of my last deployments.

Now, the calendar math is made trickier in places that use the Islamic calendar. On the Gregorian calendar—a solar calendar—the days of each month are set, but on an Islamic calendar—a lunar calendar—the length of months can be 29 or 30 days in no particular order.

I arrived at Camp Eggers in Kabul, Afghanistan, on the first day of August at the start of Ramadan, which is the Islamic holy month of fasting. Ramadan ended on the 30th of August with Eid al-Fitr, the celebration following the fast, and the new month would start the next day. However, by the Gregorian calendar, I still had an extra day or two until the start of September and the beginning of *my* second month in country.

But I went with the Islamic calendar. Even though it was the 30th of August, I told everyone that I'd just started my second month of deployment. Believe me—it makes perfect sense, even though I had been in country a short time. It is difficult to escape the count down to the end of a deployment. It can be a distraction and sometimes dangerous. In war, all thoughts that divert your attention away from the daily business of staying alive can be perilous. However, I think my slip into counting down days was more from the effect of Ramadan than anything else.

I had longed for the end of Ramadan. I really needed Eid to come. With the holiday, we would have four days to catch up on office work (and sleep) on Camp Eggers; Eid is similar to the Christmas-and-New-Year season in the West, so everyone would be off the job visiting with their families. All my Afghan friends in the Chief's office—Shuja, Dawary, and Zia—were looking forward to feasting after a month where they spent 17 straight hours of each day without food or drink. I didn't fast, but my intake of water and food was limited while at the Ministry of Defense; I wouldn't eat or drink in front of the Afghans out of respect for their sacrament and, as a result, I broke the habit of drinking water. When I did drink again back at Eggers, I would down diet soft drinks, tea, and coffee by the quart. I was still trying to get into a pattern with my sleep in a new home halfway around the world, so when the afternoon rolled around I was dozing off and needed the pick-me-up they provided.

I was also looking forward to finally eating at the Ministry with my Afghan friends and the Chief. All my fellow officers who worked in the Ministry of Defense said that Kareem, General Karimi's chef, made the most delicious meals. Once Ramadan ended, General he would prepare lunch for us and we would dine in the officer's mess. I was looking forward to it, both for the cultural and the culinary experience. The food at Camp Eggers was bland, and I ate only for sustenance. Sometimes I would bypass meals there and opt for the convenient pre-packaged foods, particularly peanut butter, nuts, and beef jerky. I'd had plenty of protein, but I was dying for some carbohydrates—and for *something* more interesting than the boiled haddock and cold broccoli the mess hall was serving.

On the last day of Ramadan, I wished everyone a happy Eid and walked back to Camp Eggers. I was thrilled at the prospect of four days for sleeping in a little, enjoying a couple good workouts, and getting a few moments of much-needed solitude.

That night, I went to bed early looking forward to a long run around the garrison in the morning. Around 10 in the evening, I got up from bed with what I thought was a back cramp. I sat up, reached around about put me on my knees. It only took a moment to realize what was wrong: kidney stones.

I had the same ailment four months before. To my horror and resignation, I knew what to expect; the pain had left deep scratches on my brain. I didn't waste any time; I got dressed and hobbled to the medical treatment center.

The last time I'd had this problem, I went to the hospital. They gave me fluids and painkillers through an IV, and within a couple hours the stones were gone. I would not be so lucky this time.

For the four-day holiday, I stayed in my room and slept; the nurse at the clinic (for whom I'm eternally grateful) gave me enough medication to tranquilize an elephant. Whenever I was awake, I drank an ocean of water. On the fourth night, the pain became excruciating even with the painkillers, so I went back to the medical facility and spent the night hooked up to an I.V. That whole fourth night, I was either curled in the fetal position on the cot or in the bathroom. I finally got to sleep, and when I awoke again, the pain had subsided a little.

Cured of the stones or not, it was time to leave. Eid had ended and I had to go back to work at the Ministry. I thanked the nurse and the medic who had watched over me and left praying that I would get through the day—because prayer is about all that could help me get through the day.

At the Ministry of Defense, everyone looked refreshed after the holiday. I ran into Kareem, the cook, in the reception area on the first floor. He reminded me of a Gucci Buddha: fat and round and smiling, but without Buddha's zen or subtlety. Kareem didn't speak a word of English so far as I knew, but I greeted him and shook his hand.

Khoob astyn? he asked. I answered *Mon Khoob astum naist*, which meant I didn't feel well; I pointed to my stomach as I said it. He replied *zin bashee*, which translates roughly to "may you be alive." I wasn't so sure I felt like it that day. He left, disappearing up the stairs to (I assumed) the kitchen.

I hung out in the foyer for a while longer and talked to the guards before climbing the two flights of stairs to the Chief's floor. The climb exhausted me; I had to stop halfway up to catch my breath. But as soon as I entered the hallway two floors above, I smelled an aroma, something vaguely familiar that had been absent the previous month. I walked into the waiting area outside General Karimi's office where Lieutenant Colonels Shuja and Dawary work.

The fragrance of chai tea overwhelmed me. One of my American co-workers, Raul Rosa, was in the corner of the room munching on an oval loaf of naan, a delightful type of flatbread. Shuja and Dawary had steaming cups of chai on their desks and were preparing to "egg knock."

I'm not familiar with all the rules, but the idea of this game is to see whose boiled egg can withstand the tap of their opponent's egg without denting the shell. The person who cracks their adversary's egg first wins.

Shuja started on the offensive. Dawary waited, holding his egg in his fist with the narrow end of the egg sticking up. Shuja blew on his own egg to prepare for attack, explaining with the utmost confidence to us that the blowing ritual hardened the egg and helped prevent cracking.

Channeling Al Pacino's Scarface (Shuja looked just like Pacino), Shuja taunted, "Say hello to my little friend." He aimed and sharply tapped Dawary's egg with his own.

No luck—Dawary's egg remained intact. Dawary's turn.

Dawary, the consummate gentleman, asked Shuja if he was ready. "Bring it on," Shuja said with brazen confidence. They stared one another down, eyes unblinking.

Suddenly, the door opened and Zia walked in. In that moment, Shuja's eyes moved from the eggs to the door, and another moment later Dawary struck. His friend's egg became a casualty of war; a small indentation was clearly visible on the shell of Shuja's egg.

Shuja claimed foul. "You cheated!" he said. But they laughed, and there were no more eggs to knock, so they peeled and ate them with the tea.

Chai and spice muffins were brought in for Zia and me. It had been the tea that I smelled coming up the stairwell; I drank a lot of chai in Iraq, but I hadn't had any since. The whiff of it brought back memories, mostly good ones. Zia told me: "Afghan tea is weak, but you will get used to it."

I took the cup and saucer. I looked down at the light yellow liquid. There were specks of brown tea grounds that swirled on the bottom. I put my lips to the edge of the cup, and I could feel already that it was hot. I closed my eyes and drank it, inhaling with my mouth and nose. It was good tea.

After I looked down from taking a sip, Zia asked if I felt okay; Raul had told him about my sickness, he said. At first the question seemed odd. Then I felt a slight pain in my back and side. I had almost forgotten about the kidney stones. *Damn it.*

I told them the irritation was dull, but still there. I told them I feared the whole thing wasn't over but that I was getting better. I silently prayed that I was right while I drank some more chai and ate the spice muffin.

The day after Eid was slow. It was full of visitors who stopped by General Karimi's office to wish him well. Very little work got done.

At exactly 12:30, the door to Chief's office opened. General Karimi came out and announced that it was time for lunch. Officers had gathered outside a room down the hall. Kareem was there, and he opened the door to a room with a long table prepared for dining. Each place setting had a large and a small plate and a soup bowl. There were two large platters of Palao—rice cooked with meat and stock and topped with fried raisins, slivered carrots, and pistachios. There were bowls of bananas, grapes, and apples. A half-dozen plates heaped with naan were spread across the table. Bottles of water and cans of soft drinks were waiting in rows.

There were no seating assignments except that the head of the table was reserved for the Chief. Everyone found a place and stood behind the chair. We sat when the boss did.

Kareem served General Karimi soup first, then the rest of us one by one, ladling the steaming broth into our bowls. I took my cue from the other Afghans and dipped the naan into the soup. It was *so* good.

There's a myth that Afghans only eat with their left hands; that myth was quickly dispelled as I watched everyone tear the bread apart with both hands. After the soup, we served ourselves rice, and I scooped far too large a serving onto my plate. Kareem then began a rotating parade of service around the table, parceling out portions of chicken, lamb, and stewed pumpkin. I was full but didn't want to stop eating. I finished the meal off with sweet green grapes.

The conversation was either in Pashtun or Dari. I couldn't tell for sure and listened closely to distinguish the sounds. The Chief peeled an apple with his knife, cut it into chunks, and then speared the morsels with his blade. He finished and stood up; we followed his lead and walked behind him as he left the dining room for his office and afternoon prayers.

I went back to the waiting room with the rest of the men. I sank into a soft leather chair and became almost comatose. I actually nodded off for a short spell, and when I opened my eyes Kareem was standing there. I woke fully and stood up to greet him and was instantly aware of the ... *absence* of pain. Kareem was smiling with his big round Buddha face like he knew something the rest of us didn't.

Khoob astyn? he asked, holding his right hand over his heart.

Khoob hastum. Tashakor, I said, thanking him as he turned to leave.

"That's good, Johnson," Kareem said. His English startled me. I turned and looked back at him. Kareem was smiling with his Buddha face.

I stopped counting down days after that. In fact, the only countdown I had was for lunch at the Chief's and Kareem's magical food.

ECHOES OF THE PAST

I dial the last few digits and wait. The automated voice on the other end says that I have 47 minutes remaining on my phone card. The phone rings three times, and Laura picks up.

Sometimes, Madelyn will answer the phone if her mom had more time when we last talked. Other occasions, it just depends on who has the most to say—and with my two girls, that can be quite a competition. I'm happy to talk to either of them first as long as I get time with them both.

We have half an hour to catch up on what's been happening, but before I know it, it's time to go. I hang up and look at my watch. Sixteen minutes left on the phone card—time to get a new one.

I walk from my office to the start point of my run near the gym. It's an easy day for me: 8 laps. I average about 6 minutes per lap: not a bad start to the morning. I finish up with a lift at the gym and head to the barracks to get ready for work.

As I'm getting dressed in my room, I glance at an old magazine I picked up in the hall. When folks redeploy, they'll put their magazines, books, and even DVDs on wooden shelves that line a wall of the barracks. It's a diversion for your fellow soldiers; during deployments, one never has to go without something new to read or watch. The magazine I picked up has an article on Albert Einstein, which I've left open; I scan it and stop at a quote in italics under his famous "mad scientist" picture (with his wild white hair and bushy mustache). The passage reads: "If I had an hour to save the world, I would spend 55 minutes defining the problem and only five minutes finding the solution."

I want to read further, but I only have about five minutes before I need to be getting to work, so I put it down and quickly shower and dress.

I walk the couple hundred meters to my office, and on the way I pick up two copies of *Stars and Stripes*, the Armed Services' overseas newspaper. The papers are for General Karimi's personal staff: LTC Shuja, LTC Dawary, and CPT Karimi, the Chief's son. It helps them with their English, and being that day's paper it usually covers important topics of conversation for that morning. For example, the week before, there was a picture of Al Pacino in the paper and we had a lengthy debate over whether Shuja looked more like Scarface (Pacino in the 80s) or Michael Corleone (Pacino in the 70s). Shuja thought he looked more like Michael Corleone; everyone else thought he looked more like Scarface.

I have an hour before I have to leave for the Ministry of Defense. I print off the most current news items for the General, finish up my notes, and talk through a couple minor issues with the other team members in the office.

At 0700 on the nose, Tom Williams enters the room. Tom is a 65-year-old retired Marine and Vietnam vet. Tom is fluent in at least four languages (that I've seen him speak); he has spent the last five years throughout the Middle East and Southwest Asia working for the government. When he's not deployed, he lives with his Rhodesian wife in Australia; I've not met her, but I'm sure she's impressive. Tom knows only excellence, and he has the distinction of being probably the most unique person I've ever met.

But Tom is also a joy to tease. It's so much fun that I've made taunting Tom a part of my daily routine. I usually poke fun at the Marine Corps or joke about his remarkable resemblance to Mr. Magoo (albeit a leaner and far more articulate version). Today, I choose a highbrow approach and quote Rudyard Kipling. "It's Tommy this and Tommy that ... and Tommy *go away*," I said, pretending to sneer.

Tom finishes the stanza. "But it's 'thank you Mr. Atkins' when the band begins to play," he said, smiling.

He returns fire by reciting from my Kipling favorite. "If you can fill the unforgiving minute with sixty seconds' worth of distance run, yours is the Earth and everything that's in it, and—which is more—you'll be a man, my son!"

He chuckles. Then he puts his hand on my shoulder and starts cutting. He says, "On the subject, I saw you running today. It looked like you were filling that *minute* with about 30 seconds' worth of run."

He removes his hand from my shoulder. "Are you putting on weight? You're awfully slow for such a young man."

I laugh as he turns to leave. He slides his hands into his pockets, jingles the change in them, and saunters away whistling the Marine Hymn. *Touché, Tom—one for the Corps.* I pack up my things and begin my trek to the Chief's office.

Today, we are flying to Herat, a city in the westernmost province of Afghanistan, not far from Iran. The Greek historian Herodotus called this 2700-year-old metropolis "the bread basket of Central Asia." It is a beautiful city and was supposedly one of Alexander the Great's favorite places to stay as he marched to the end of the known world. Herat has been spared much of the destruction visited upon other parts of the country.

We are going to a couple of bases on the outskirts of Herat to check on soldiers there and inspect the construction of some new facilities at Shindand Air Base.

Before getting on the plane, I give General Karimi a packet of reading materials that include relevant newspaper articles and a packet of information that highlights key issues for him to review. One of the news pieces is a recent *Wall Street Journal* piece about the Afghan National Military Hospital.

A year ago, the Coalition uncovered criminal neglect and horrendous care of wounded soldiers at the hospital. Amputations were being conducted without antiseptic and painkillers. The hospital's medications were being sold on the black market. The neglect was so bad in this facility that one highly-decorated soldier actually died of malnutrition. The situation reminded me, in a way, of the U.S. military's scandal at Walter Reed several years before. Our Army took steps to fix the problem; likewise, General Karimi and other leaders in the Afghan Defense Ministry made changes and treatment got better. (Of course, the journalist who wrote the article failed to mention that the incident had taken place a year before and that many improvements had been made since.)

Karimi says he's already read the article, so he turns past it to the information packet. When I first started compiling the information, it only had a few list items, maybe a half dozen; now, it's six pages long.

In fairness to the Afghan National Army, if you compared them against any other armed service, you would find that they have many of the same problems; the difference is just a matter of degree. Still, it's those matters of degree that can make for such difficult problems.

One of the biggest obstacles to building the Afghan Army is the high rate of non-combat-related attrition—people leaving the Army for reasons that have nothing to do with the battles themselves. It is bad enough that their Army is losing soldiers to battle injuries, but that is normal for an army at war and ultimately can't be helped. It's a different kind of problem when large numbers of soldiers are going AWOL (absent without leave) and many more are opting not to re-enlist. This attrition is happening at such a rate that the Afghan Army is losing a quarter of its people every year, which is incredibly costly.

As General Karimi reads each data point in the report, his shoulders gradually slouch, like weights are being stacked on them one metal plate at a time. He keeps reading, but before the burden breaks his back, he closes the folder and puts it away. He shakes his head. "Poor leadership," he says. "That's the problem."

Our flight from Kabul to Herat is 470 miles, about an hour of flight time. General Allen, commander of the International Security Assistance Force, let my boss, Lieutenant General Bill Caldwell, use his jet. During the trip, Caldwell's and Karimi's seats face one another, and they talk about the state of the Army and the approaching U.S. drawdown in forces.

By 2014, the combat mission for the Coalition Forces will be over. This year, we will reduce our force by 10,000 soldiers, down from the surge authorized by President Obama two years ago. By the end of next year, we will remove an additional 23,000 service members from the region. General Karimi and most of the Afghan commanders have grave concerns about how they will fill the gap left by the Coalition. Karimi reminds Caldwell that "this is a very difficult problem."

Our first stop is at a regional training center where the Army conducts basic training for new Afghan recruits. After we've landed, I look out over the field where Afghan soldiers are practicing artillery drills with large Russian-made howitzers. The Coalition trainers—senior enlisted and officers from Italy, Canada, Australia, and the U.S.—hover on the periphery while the Afghan trainers lead the Afghan soldiers through the exercises. That's the way it's supposed to be, but it's not the way it's always been. I can tell General Karimi is satisfied that his trainers are in the lead.

I stand back with the other Coalition officers as Karimi moves inside the circle of soldiers and trainers. I have seen this move before with American generals; whether or not they can pull it off tells you a lot. For some senior officers who enter the circle of enlisted men, it is clear that they are in a foreign and unnatural place where they aren't comfortable. Then there are officers who blend into the soldiers as well as the soldiers blend into one another.

General Karimi enters the ring of servicemen and disappears. I can still hear him speaking Dari, and I ask an interpreter nearby to translate for me.

Karimi says, "You must know every part of the equipment you use. This artillery is like a person. You must know the head, the body, the hands, and the feet. Each part has a purpose and you must know that purpose perfectly." The pitch of his voice raises and lowers; the heads of the soldiers bob with the inflection of his tone, gently hypnotized.

We re-board General Allen's jet and take a short 10-minute flight to Shindband Air Base. The Afghan Air Force is very small; at this point there are only 51 fixed-wing planes and helicopters. There are no fighter jets, only transportation aircraft, several of which are reserved to transport dead soldiers. (In Afghanistan, moving dead soldiers often takes priority over evacuating the wounded. The Muslim funeral ritual requires that the dead be buried with within hours, but no more than a day, after death. Some Westerners are appalled by this notion, but it's not a debate worth having in an Islamic country where one's place in paradise and the everlasting is more important than their earthly existence.)

While the Air Force is still small, Shindband is huge and growing. The U.S. is spending $280 million on the base to build what we call "Afghan Right" facilities. It took almost ten years for us to learn to construct buildings with amenities consistent with Afghan culture and tradition—otherwise, they don't get used as you intend. For example, toilets were being broken because Afghans were standing on the seats to relieve themselves rather than sit on them. Urinals were being used for bathing before prayer. With often good intentions, Western nations built facilities that were consistent with customs of Western culture. This was a very costly and time consuming problem, and the solution had only recently been realized with "Afghan Right."

Shindand is so big now that we have to take a helicopter to see the full magnitude of the effort. As we tour the base from the air, I notice what looks like an enormous junkyard in the distance; as we get closer, I can pick out dozens of Russian MIG fighter jets and hundreds of former Soviet tanks and armored vehicles piled atop one another. The rusting pile of metal and scraps has to be a half a mile wide, at least.

I ask General Caldwell's planner, Colonel Dan Klippstein, if they were destroyed by the Mujahedeen during the Soviet occupation in the 1980s. He shakes his head no.

He points to the MIGs and says, "The Soviets gave those to the Afghans in the 70s. 'State-of-the-art' equipment."

He can tell I'm having difficulty hearing him over the helicopter. Dan cuffs his hand over his mouth, moves closer, and shouts into my ear so I can listen. "They gave them the gear but didn't teach them how to maintain it. In a couple years, planes, tanks, artillery pieces, they all started falling apart. There are dumps like this all over the country. Dozens of them. Hundreds of millions of dollars just wasted."

"It's called 'Echoes of the Past'," he adds before turning to look out the window somewhere else in the distance.

We fly over the far northern end of the Echoes, passing a MIG with its nose up as though it's ready for takeoff, except the cockpit windows and landing gear are missing. We slow and hover over new construction that will eventually be troop barracks.

Dan pulls away from the window, puts his hands on his knees, and turns back toward me. Still shouting over the noise, he says, "They didn't teach them how to do it for themselves. That was the problem."

THE VISITOR

Last week, General Karimi and I were driving back from a meeting. It was late afternoon; we were taking the long way back from Afghan Logistics Command. We never take the same route to and from a location. Karimi's son, Zia, is very meticulous in planning our movements.

Around 5 o'clock, we passed an Army checkpoint. General Karimi ordered the driver to stop. We got out of the armored Chevy Suburban and the Chief stepped out and walked into the guard shack. His security detachment surrounded the door. I asked Zia, "What's up?"

Zia pointed to the sun, then to his watch. "It's prayer time," he whispered.

I should have known. General Karimi never misses his prayers. He will skip a meal before he fails to perform Salah.

Muslims pray five times a day: before sunrise, after lunch, before sunset, after dinner, and then before bed. General Karimi once told me that his prayers are the only time he truly feels at peace. He said that, when the Soviets put him in prison after the invasion, his prayers kept him together. Every so often, the guards would take him to the prison courtyard and put him through a mock execution. Then they would say, "Maybe we will kill you tomorrow."

Karimi told me later, "I say my prayers every day. So I am always prepared to die."

Today, I walk into General Karimi's waiting room and acknowledge the staff with one of the two Pashto phrases I can pronounce. *Tha tsanga ye?* I say, and they reply with *jouri*, the other expression I've learned. Pashto is incredibly difficult, and Zia warned me to concentrate on Dari first, but many of the Chief's guards are Pashtun, and when I practice my Dari on them they tell me, "No Persian. Talk Pashto."

While both Dari and Pashto are Afghanistan's official languages, there is a very strong cultural, historical, and even religious divide between the people who choose to speak one tongue or the other. Dari, called Afghan Persian, is considered the cultured (highbrow) literary language. Pashto is the spoken by the tribal Pashtuns as their native tongue.

When I speak Dari to the guards, they become like missionaries bent on Christian conversion of a sinful savage. They very patiently walk me through their tremendously difficult dialect. I understand more and more why Zia says it's hard; Pashto has 12 more letters in its alphabet than English and a collection of throaty, tongue-rolling sounds that native English speakers often can't make well.

I'm uncertain whether the guards' persistence in teaching me Pashto is altruistic or just for their own entertainment; inevitably I tangle the words and raucous, childlike laughter ensues. Either way, I indulge them; they protect the office, the Chief, and ultimately me.

Even if it's all a joke to them, I consider it a fair trade to be the butt of their jokes.

Before going into Karimi's outer office, I notice two men on one of the couches in the lounge. I've never seen them before. I put my right hand on my chest next to my heart and say *Sa-laam Alaykum*. Neither responds. They stare straight ahead, sitting motionless like two statues of Afghanistan's archetypical rural populace.

One man is young; the other appears ancient.

The older man's face looks like the Hindu Kush Mountains, only upside down. His unkempt white beard is a broad downward summit that covers his neck, the whiskers gradually receding into deep valleys and dark ridgelines on his weathered face. In all of that harsh, age-beaten skin, the only soft spots are his brown eyes; I might have said they were kind eyes, except his lack of acknowledgement of my greeting betrays that inclination. His turban and cane also give me pause for concern, even though they are perfectly normal in Afghan culture.

It is no secret that the enemy, particularly the Haqanni Network, is targeting senior Afghan civilians and military leaders in spectacular attacks to gain media attention. Former Afghan President Burhanuddin Rabbani was killed recently by a suicide bomber who hid explosives in a skull cap under a turban. The killer gained Rabanni's trust and waited four days to conduct the attack. He detonated himself while hugging the Afghan president and the standing chairman of the High Peace Council.

My spidey senses are tingling out of control. Are the old man's turban and the wooden cane assassin's tools?

It wouldn't be the first time someone had tried to kill General Karimi. In April of this year, a man entered the Ministry dressed as a soldier. As the guards stood to greet him, the man turned his weapon on them and shot them dead. The shooter worked his way from floor to floor towards the Chief's office.

Shuja heard the shots. He quickly organized Karimi's security detachment in the hallway outside the stairwell—the only entrance to the floor. As the assailant turned the corner, Shuja and the guards opened fire. As the bullets punctured the attacker, he grabbed at his sleeve as though reaching for something. After what Shuja calls the longest five seconds of his life, the man finally fell dead. When the soldiers searched him, they found he was strapped with enough explosives to bring down the whole building. It was a miracle the shots didn't detonate the explosives. The man was grasping for the detonator of his suicide vest when Shuja put one final bullet in his head.

I walk into Karimi's outer office. Shuja and Lieutenant Colonel Dawari are there. I ask them if they know the old man on the couch and they shake their heads no. I tell them about the cane, turban, and his unfriendly demeanor; I remind them of the details of Rabanni's assassination. Shuja goes out and talks to the old man. Shuja talks to him in Pashto, very politely, with the reverence that is expected to be given an elder. They speak for a while, and Shuja asks to see his cane. He looks at it as if it were an art piece to be admired. He shakes it gently as if to appreciate its sturdiness and then gives in back to the man. Shuja comes back and says the man is here to see the Chief. "The cane seemed fine, nothing unusual." Dawari informs General Karimi he has a visitor who isn't on the calendar, and Karimi says he is aware of the appointment and tells Dawari to show them in. Both guests go into the office without the cane.

Within seconds of the door closing we hear raised voices and indiscernible noises. Moments of quiet are interspersed with clatter. It's unclear what's going on. After a few minutes, the door finally opens. General Karimi and the older man come out, followed by the younger fellow. All three are smiling, speaking rapidly in Pashto, sometimes over one another in a warm exchange between men who are clearly good friends. I figure it out: the noise was laughter.

Karimi says in English, "This is my classmate and good friend from Khost (the Chief's home town in the mountains of eastern Afghanistan, an area strongly Pashtun). We served in the old Army together until the Soviet occupation. He was a great Mujahedeen and fought the Russians."

Karimi laughs. "I went to jail; he went on Jihad," he says of his friend. Then he says, "Let's have lunch."

We follow them out, and behind their backs Shuja mimes wiping sweat from his brow to me.

Before following them to the dining area I return to the waiting room and pick up the cane. It's lighter than it looks. The thought crosses my mind that it's hollow. I carefully study both ends, then I put it back down and join everyone for lunch.

THE QUEEN'S PALACE

★ ★ ★ ★ ★

My favorite high school teacher was Ms. Polly Peterson. She had also been my parents' favorite teacher. I asked my dad once if Ms. Peterson was pretty when she taught him, and he replied that "Polly was old even then, but I heard she was a looker in her own day."

If I polled my 1980 high school class and asked, "What is the first book you ever read from start to finish?" I bet 90 percent of the graduates would answer *Great Expectations* by Charles Dickens. It was required reading in Ms. Peterson's freshman course. Her class was one of the best things ever to happen to me, and it incited my adulthood love of reading and writing.

I always felt there was some irony that Ms. Peterson made us read the Dickens classic. When reading about Ms. Havisham, one of the central characters of *Great Expectations*, I simply could not help but think of Polly. Ms. Havisham was an heiress to a fortune, but she was left at the altar on her wedding day and scorned to the point of madness. She lived in a dilapidated mansion called the Satis House where all the clocks stood still at 20 minutes to nine. The Satis House was a metaphor for Ms. Havisham's decrepit soul, ravaged by unrequited love.

I'm not sure why Ms. Peterson reminded me of Dickens's spinster. Polly was a wonderful person and very well-kept, *elegant* I would say now. She was petite with short, cropped white hair and deep wrinkles. Her red lipstick complemented the variety of black dresses she always wore. But to a hormone-riddled 14-year-old boy, she seemed ancient. For whatever reason, Polly Peterson and Ms. Havisham will forever be joined in my mind.

And it was Ms. Peterson and the Satis House that first came to my mind when we entered the grounds of Camp Julian, where I would be training for the next few days.

We had taken the Rhino from Camp Eggers to Julian. "Rhino" is the unfortunate name for an armored bus that transports soldiers, Department of the Army civilians, and contractors to secured areas within combat zones. It is not a fighting vehicle; rather, it is protected by two other armored trucks with heavy weapons, one in the front and one in the rear.

While the Rhino is a daunting, mobile fortress, that often lures its passengers into a serene sense of safety, it is also a lucrative target. My training was supposed to have started several days ago, but it was delayed because a Rhino carrying a dozen occupants was attacked by a vehicle loaded with explosives. Four people were killed. The attack took place right outside Julian where we were scheduled to enter the camp.

I was crammed in the Rhino with 20 other people. I felt utterly ridiculous peeping through the thick reinforced glass while women and kids shopped at local markets. The three massive vehicles snaked at a snail's pace along narrow roads. However strange they might have looked, the passersby were inoculated to the presence of the massive military vehicles; young boys raced across our path to get to the other side of the street, unconcerned they might be smashed like a bug under the wheels. Uniformed policeman had to stop traffic while we pushed our way through tight intersections until we finally turned into the entrance of the camp.

Passing through the gates of Camp Julian (very relieved we hadn't been blown up), I noticed a large structure on a hill. At first, I thought it might have been an old government building, partially destroyed during the 30 years of war and left to crumble. But as we got closer, the odd shape became more ornate, and in fact the hill itself had been sculpted to accentuate what rested above. There were several walled tiers leading to the top, where the massive edifice watched over all of Kabul. The building itself had more windows than I could count and there were splendid balconies on each end. It was simply magnificent even in its state of disrepair.

I've come to Julian to receive the final days of training in a program designed to teach us how to advise senior-level leaders. I've been on the job for almost two months and am finding the instruction fascinating and helpful—but in ways I'm sure the course designers did not intend.

In one of my favorite classes, the instructor wears a *pawkul*, the traditional Afghan headgear, and he has a *keffiyeh* scarf wrapped around his neck. He is tall, bearded, and animated; I like him immediately. I sit in the front row, and to impress him I greet him with my very best Dari *good morning*. He replies with a crisp American accent: "Hey man, what's up?"

I thought he was an Afghan. But, to my surprise, I learn he is a retired Army officer with a Ph.D. and a backpack full of opinions on why we are losing the war.

The purpose of his instruction is to teach us how best to fight a counterinsurgency, as though groups like the Taliban and Al Qaeda were the entire problem. But I come away from the instruction thinking that the greatest threat to security in Afghanistan is not the insurgents at all. In moments of distraction, while discussing the intricacies of Afghanistan's culture, he says just as much about its internal problems, especially corruption.

Patronage networks within high levels of military and government leadership place people in positions of power based on ethnicity and personal favor: it's nepotism that weakens a framework that might otherwise create a talent-based meritocracy. Illiterate fools make decisions with strategic consequences, decisions that put lives on the line. Corruption erodes the rule of law to the point that even a draconian system of Sharia law can't or won't address it. Not only is its government corrupt, but Afghanistan is divided between the Tajik-led Northern Alliance and the Pashtun majority in the South—their very own Mason-Dixon Line, which threatens the very sovereignty of their nation as ours once did. These weaknesses are made worse by the Taliban's safe haven within Pakistan that allows them to attack and erode Afghanistan from the outside.

We are engaged with an array of declared enemies in Afghanistan, and most are aligned in one way or another with the Taliban. The instructor argues that the strength of Al Qaeda, the reason we came to the country, has been diminished considerably. I wonder whether that really makes a difference in the long run.

The other class is on "the Queen's Palace," which I learn is the name of the manor on the hill. An Afghan-American doctoral student gives us a brief on the history of the site. The instructor's dissertation was based on his interpretation of the graffiti left on the remains; Pakistanis, Soviets, and every variety of Afghan, along with Americans, Canadians, and other coalition soldiers, left their personal marks in nearly every room. The walls almost literally talked, and he'd wanted to tell their story.

From his lesson, I learn that, after Afghanistan won their independence from Great Britain in 1919, the king Amanullah Khan built the palace complex for his wife Soraya Tarzi. King Amanullah was known as a great reformer; he built schools for both boys and girls, increased trade with Europe and India, and wrote a constitution that significantly increased civil rights for all Afghan citizens.

In 1927, soon after the palace was completed, the king and queen had a grand ball for dignitaries and tribal elders. At the gala, Soraya, in an attempt to jumpstart the suffrage movement in Afghanistan, unveiled herself. This was only six years after women in the U.S. had gained the right to vote.

Two years later, King Amanullah and Queen Soraya were forced into exile by religious conservatives.

The Queen's Palace and its adjacent structures, which were intended to be the seat of Afghanistan's government, were left deserted for many years. Finally, three decades of war left the entire compound virtually in ruins.

After the lecture, we walk the half-mile up the road that wraps around the hill to the entrance of the palace. Once we get there, we are allowed to explore on our own.

I recently received a care package from my good friend Dr. Beth Johnson and was overjoyed to find a box of exquisite Imperator cigars inside; I've brought one with me in the hope I will find a place worthy of enjoying it. I take it from my pant leg pocket and place it in mouth. I choose to keep it unlit for now and chew the end gently with my teeth. The leaves become wet and softened, releasing the soothing flavor of tobacco onto my tongue.

Cigar still in my teeth, I start on the first floor and walk around until I work my way up to the third floor. We are not allowed to venture any higher because the upper floors are no longer weight-bearing. Where the walls are still intact and not wasting away, I try to read the graffiti. Most of the writing is in Dari, Pashtu, or Pakistani Urdu script that I can't make out, but there are more than a few passages in English from American and Canadian Soldiers, mostly bravado about the unit to which they were assigned. There are disparaging comments from some, about their current station in life thousands of miles from home, but nothing obscene or surprising.

When I've explored as far as I can on the third floor, I meander from room to room looking for the best vantage point of Kabul so I can take a few pictures. I find the best spot on one of the middle balconies of the house; it provides a panoramic view of the city with the mountains in the background.

After shooting a few photos, I take a matchbook from my pocket and light my cigar, sucking in deeply to pull oxygen through the flame and ignite the tobacco. I breath out the rich smoke and watch it float out the window and fade into the air. I sit on the ledge of the veranda to get a better angle of a mosque in the distance. I shift my weight, and as I do I loosen a chunk of cement that falls to the floor.

The Queen had such great expectations, I thought. This place could have been alive—could still have been alive. But I'd lived in a time when I could not know its beauty—and then it was only an empty place, where the walls were covered with messages that would no longer move in time. A piece of that wall has come down beside me.

I pick up the concrete and it crumbles in my hands.

SLOW IS SMOOTH, SMOOTH IS FAST

Before I deployed to Afghanistan, I hadn't shot a pistol in over three years since I was stationed at Fort Lewis, Washington with the 3-2 Stryker Brigade and my last tour in Iraq. As a part of pre-deployment training at Fort Benning, Georgia before leaving, I had to qualify with my 9mm. The pistol felt like something completely new. I *remembered* how to use it, the way one knew how to ride a bike after years of not riding one—but I felt uneasy about shooting well, the way you might feel uneasy about riding a bike for the first time in several years.

Fortunately, they gave us a practice round to regain familiarity with the weapon before the actual qualification.

My practice set was abysmal. I rushed my shots on the pop-up targets. I had to think about turning off the safety and releasing the magazine. I searched with my hand for my next magazine instead of instinctively grabbing one for reload. Everything was forced, everything was hurried. It is the same as if I were told to shoot a jump shot after not playing for several years. It is a skill that atrophies if not practiced regularly. I was woefully, and embarrassingly, out of practice.

Then, before starting the qualification round, I remembered a saying we'd learned while rehearsing close-quarter battle. It is the infantryman's proverb of *Slow is smooth, smooth is fast.* In moments of stress, there is a tendency to rush, and when you rush you miss important details, which actually slows down the process (and possibly gets you killed). However, if you mentally and physically slow down, the action is smooth, no details are missed, and the activity is completed efficiently. It is also called *being in a careful hurry.*

Relax, I said to myself. *Slow down.*

I would love to say I shot expertly; I didn't. I shot well enough. More importantly, though, I regained my confidence.

I think of that occasion at the range every time I leave Camp Eggers with a weapon. I pass through the wall of the compound by walking between two reinforced steel doors; one shuts before the other opens. Before leaving completely, I chamber a round into my weapon. I smile at the guard and say, *Sa-laam Alaykum.* He responds, *Walaykum Assalaam.* I holster my weapon as I pass to the other side, and I consider the irony of the Islamic peace salutation in that situation, where I've just locked and loaded for war.

I work in the Green Zone, but the April assassination attempt on General Karimi at the Ministry of Defense, along with a recent attack on the British Consulate, remain fresh in my mind. Just a week ago, rockets exploded near President Karzai's palace.

I recall the advice of Brigadier General Steve Townsend, my brigade commander in Iraq. He sent me an email upon his redeployment after a year here as the deputy commander of the 101st Airborne Division Air Assault. Repeating one of his favorite sayings, he wrote, "Keep your pistol clean and loaded. Stay on Amber, Fred."

Amber is shorthand for the mental place between hypervigilance ("Red") and normal, day-to-day calm ("Green"). One cannot stay overly alert, or "Red," for extended periods of time without burning out. On the other hand, the mindless tranquility of "Green" can get you killed when you're not looking. But someone can stay "on Amber" indefinitely.

Today we are going to Qargha, which is just outside of Kabul. Qargha is where the United States and Great Britain are building the Afghan National Security University. All the Afghan military education institutions—to include Sandhurst (their version of West Point) and the War College—will be in this one central location at Qargha. General Karimi has to conduct an inspection there in preparation for a visit by senior-ranking dignitaries.

Before leaving for Qargha, I have my daily meeting with Karimi. There are a couple items to discuss, mostly pertaining to units that are tardy in meeting deadlines and organizations that have training and logistics shortfalls. I cringe when I bring the points to his attention. On our first day together, Karimi had said to me, "All I ask of you is to have tolerance and patience." Having spent several years training and going to school in the United States, he knows the lack of Western restraint all too well.

I am hesitant to push too far or too fast. I did in this in Iraq trying to re-establish an oil refinery near the city of Mosul. I was in such a hurry to get the refinery producing gas that I failed to ensure certain important details were in place first, like having skilled technicians willing to run the facility. As a result, our progress was set back months and we never fully recovered. The refinery never became fully operational. I had *rushed to failure.*

I pass General Karimi a couple information papers I wrote on the topics for his consideration. I'll give him a couple days to reflect, and then I hope to address them again. Done with the meeting, we are now ready to head to Qargha. I'm excited to get out and see the countryside.

Our convoy takes off from the headquarters, and I ride with my most recent Facebook friend, Captain Zia Karimi, General Karimi's son. Zia just recently returned from studying in Italy and now helps coordinate security for his dad. Zia is a striking young man of 23; he likes to say he looks like Sam Worthington, the movie star who played Jake Sully, the paraplegic Marine in the popular movie *Avatar.* One day my assistant, Raul Rosa, brought in a picture of Sully's Na'vi avatar—basically his blue alien counterpart—and said, "Is this who you mean?"

"That's . . . wrong," was all Zia could say.

Our convoy leaves the gate and the rollercoaster ride begins. Driving lanes do not exist in Afghanistan—not for cars going in the same direction, not even for cars going in opposite directions. Even though we have a military escort with well-armed soldiers in the lead and rear, Zia is constantly dodging oncoming traffic. He must have had defensive driver training; I wouldn't have made it a block if I were behind the wheel. Zia sees that I'm somewhat uncomfortable and turns on the radio. He says, "Do you want me to find you a Western station to listen to?"

My eyes widen at a near-head-on collision in front of us and I timidly tell him, "No, it's okay."

He puts in a CD. "It's music from India. It's relaxing music," Zia says calmly, narrowly avoiding a motorcyclist.

I am convinced it is only through the great skill of Zia and the other drivers that we get to Qargha alive. I've sweated completely through my Multi-Cam uniform even though the air conditioner has been on high the entire way. But seeing Qargha is worth the drive and the seven near-heart attacks it caused me. The Military University is situated in a wide open space between three mountains. Zia tells me there is a pristine lake just on the other side of one of the mountains where people go to picnic. He says it is one of the most beautiful places in Kabul. I can think of no better location to educate and train the future leaders of Afghanistan.

A British colonel gives us a tour of the National Military Academy grounds, the Afghans' future West Point. The garrison is nearly complete, and they expect to be done in six months. We look at the male barracks, then go through the female cadet quarters. (The mere thought of women attending Afghanistan's military academy boggles my mind.) Then we go through the headquarters building, dining facility, and auditorium. With the tour complete, General Karimi is clearly moved by the quality of work that has been done, and he says, "This is the future of my country."

General Karimi has an empty seat in his vehicle for the trip back, and he asks me to ride with him. As I open the door to get in, Zia looks at me as if to say, *chicken*. For the ride back I allow my mental threat level to momentarily shift to a somewhat lighter shade, closer to yellow than orange.

We pass ladies in burkas shopping at roadside markets, and I think of the female barracks at the National Military Academy. Soon young women will parade side by side with their male compatriots.

The first female cadets entered West Point in 1976.

I see young boys who are Hazara, a minority ethnicity in Afghanistan, playing in a park. I recently met Lieutenant General Maroud Ali, the newly appointed Army Ground Force Commander, himself a Hazara.

It was 1948 when President Truman signed the executive order to racially integrate the Armed Forces. In 1954, Benjamin O. Harris became the first African-American to be promoted to the rank of brigadier general.

I reflect on the role of the NATO Training Mission in Afghanistan, the organization to which I am assigned. I think of how General Karimi politely refers to us as his mentors. He is always humble and effusive in his praise of our support. He says, "We have made a lot of progress in the last ten years. But it takes time to build an Army."

I am reminded of our own nation's struggle more than 200 years ago when our forefathers were challenged to create an Army *while* fighting a war. In that conflict, we asked a foreigner—the Prussian Baron Von Steuben—to lend a hand in training and transforming raw, ill-disciplined recruits into soldiers that could to do battle against the British, one of the finest militaries in the world.

We near the gate to Camp Eggers and General Karimi asks the driver to slow down. He asks me if I want to be dropped off or go back to the headquarters. He inquires if there is something else we need to discuss.

There are a couple pressing issues and my superiors want immediate answers. I should probably go back with Karimi and discuss them.

I consider it for a second.

I say to the Chief, "I'll get out of here, sir. See you tomorrow."

STRONG WOMEN

I love Thanksgiving. It is, by far, the best deployment holiday. The dining facilities go to great lengths to prepare a feast comparable (almost) to anything my mom made when I was growing up. On this Thanksgiving morning, I get up dreaming of turkey, mashed potatoes, and stuffing. I call home to talk to my wife, Laura, and daughter, Madelyn. Thanksgiving is a very special family holiday for us. Beyond the meaning of the day—and, of course, the food—it marks the anniversary of my and Laura's engagement.

We had met two weeks before Thanksgiving in 1996 at Fort Leavenworth, Kansas. I had just returned from Bosnia and was literally living in my office until I could find an apartment. My after-work routine was to go to the gym, work out, shower, and then have dinner and beers at a pub on post called the Havana Beach Club. Each time I went to the gym I would see this gorgeous woman just killing it on the Stairmaster exercise machine. She always wore baggy Umbro soccer shorts and a cotton race T-shirt over her sports bra. Sweat gathered in pools on the floor around her to the point people would broadly circumvent the exercise machine she was using to avoid being rained on by her perspiration. I later found out she was a captain, like me at the time, and a military police officer. I fell immediately in love.

I never gathered the nerve to talk to her, mostly because I was afraid I might drown in her sweat. However, one night she was at the Havana Beach Club having a beer with her senior non-commissioned officers. My moment had arrived. A buddy of mine asked Laura and one of her friends if they wanted to play doubles pool. They accepted, and Laura was my partner. As we played, I brazenly started giving Laura tips on how to shoot pool, oblivious to the fact that I was annoying her. We were shooting 8-ball, and it was very evident that I wanted to win badly. I was coaching her to cut a ball in the corner pocket. She looked at me and in a contrived dense-blonde voice she said, "You know, black is my favorite color. I want to hit that pretty ball right there." Before I could say anything, she sank the 8 ball and we lost the game.

Laura then proceeded to run the table of the remaining balls. As the last ball crept toward a side pocket and fell, she threw her pool stick on the table and said, "Don't you ever talk down to me again, Sporto." I was now out of my mind in love.

I somehow recovered from that incident, and we started dating. Most of those dates involved working out at the gym or running the hills of Fort Leavenworth. Inevitably, we'd end up at the Havana Beach Club, where she would school me in pool. Those two weeks were a whirlwind. I then asked her to come home with me to Illinois and have Thanksgiving dinner with my mom and dad. She agreed. After that meal, we announced we were getting married. The next day we bought simple gold bands for one another. On Christmas Eve, a month later, we were married at my house in Illinois by a Pentecostal preacher who was a childhood friend.

Our six-week courtship turned into a 15-year marriage to the most remarkable woman I could ever imagine. Laura got out of the Army, earned her Ph.D. and became the director of behavioral health at Fort Knox, Kentucky. She leads 80 psychiatrists, psychologists, and social workers in the incredibly important missions of maintaining the mental health of soldiers, civilians, and veterans and specifically treating service members with PTSD.

Over the phone this Thanksgiving we recount our love story to Madelyn, who has heard it dozens of times already. We all laugh and Madelyn comments, "Mom, you were a real badass." Madelyn is already starting to resemble her mother in that way. She is becoming a very good athlete, but more than anything, she has inherited Laura's strong will, which is the trait I admire most in both of them.

Our call comes to an end, and we say our goodbyes. I hang up the phone and get ready for my day's work.

Today General Karimi and I are going to the graduation ceremony for one of the first female Officer Candidate School classes in the Afghan National Army's history. Karimi will preside over the ceremony so I move to a location away from the crowd. I find a ledge of a walkway to the headquarters building of Kabul Military Training Center, which stands above the procession grounds. As the dignitaries gather and greet Karimi I can see the soon-to-be commissioned female officers in a distance as they prepare for the historical event.

Ten years ago Afghan women were seldom allowed outside the home, and when they were they were required to wear full burqas. They were provided only a basic education and college was out of the question. I thought of Barack Obama's presidential election in 2008. I recalled my pride in the symbolism of an African-American's selection to our nation's highest office and the belief that prejudice and discrimination had finally been placed in the dustpan of our history. Similarly, the commissioning of these women as officers in the Afghan Army leaves me with such an overwhelming sense of elation that I want to experience it alone as I ponder the significance of what I am about to witness.

As the band warms up and the audience take their seats I continue to watch the Afghan women. Their look of determination strikes me and Julie Walter comes to mind.

It was 2007 at the height of the surge in Iraq. Command Sergeant Major Julie Walter signaled me from the doorway of the Food Ministry in Baghdad's Sadr City. It was August and CSM Walter's uniform was soaked. Perspiration dripped from her nose, ears, and chin. Her body armor and neck guard were stained from the combination of sweat and Iraq's dust and grit. You could barely discern the digital camouflage pattern on her uniform. Her vehicle, which was right outside the Ministry, had just been attacked by grenades. I was in the middle of a pretty tense situation myself. Security guards had surrounded our small team. They were intent on kicking us out of the Ministry before we could finish our mission. I broke free and met her at the door.

CSM Walter had her M4 at the low ready, a non-threatening posture that oriented the muzzle of the weapon at the ground, but it was held in such a way she could quickly raise it and engage a target. She stepped into the building and nodded over to the angry security guards and said, "Why are they being dickheads?" I told her they're just trying to flex and intimidate us. I changed the subject and asked, "What's the status outside with the soldiers and vehicle?" "The grenades landed just outside the vehicle. Shook some folks up, but everyone is okay. No worries," she said.

Julie Walter was from Portland, Oregon, and had joined the Army right out of high school. She had high cheekbones that would make you think she was Native American if it were not for her skin tone that reflected more Irish stock than anything else. If we were not soldiers, and particularly close teammates, I think I would consider her pretty. I had never given that a thought though. Regardless, I could imagine that her ancestors were a part of the wagon trains that crossed America and settled the west. Julie had a strength that came from a bloodline of survivors. There was no one I trusted more.

CSM Walter shifted her shoulders to adjust the weight of her body armor and looked over to the security guards. She paused for a second and quickly licked her lips to salve the deep cracks from Iraq's sun. She said, "Alright now, sir, finish up with the frat boys over there and get the food so we can get the heck back to Baqubah." I hesitated and before I could speak she said, "I got everything under control outside. The soldiers will be just fine." She then opened the door and stepped out into the oppressive heat and the soldiers that awaited her leadership.

The sound of the band preparing to the play Afghanistan's national anthem snaps me back from my reverie and my attention is directed to the ceremony that is about to take place. I got to attention and salute until the final notes fade. I drop my salute and look to where 14 young women, wearing their dress uniforms, stand in formation around a long rectangular table which holds a Qur'an and an M16 rifle for each of them, lined up side by side. Under their formal military headgear, their heads are covered with black scarves, the only visible difference between them and any other group of Western female officers.

The class leader stands at the head of the table. Later in the day, she will earn an on-the-spot promotion to First Lieutenant from General Karimi for her exceptional performance during training. The young woman is a Hazara, an Afghan minority ethnicity that has often been persecuted, particularly by the Taliban. I move forward to very edge of the walkway to get a better look. I want to get a good view of her. She has Asian features, smooth skin, and dark eyes. There is no physical feature that would have compelled me to think of Julie Walter—none except the inexplicable strength that is inside the eyes.

She shouts a command and the group snaps to attention. With another crisp order, each of them places one hand on the Qur'an and the other on the rifle. The women's heads are rigid, turned toward the leader. Their eyes are set on her as she reads and they repeats their oath of office. With the oath complete, the Afghan National Army, now has its newest officers. They are joining a military where less than 10 percent of the personnel are female, compared with about 15 percent in the United States.

I watch the officers march off to an area where their loved ones and fellow soldiers meet them. They show off their insignia that designates them second lieutenants. They strike poses with their rifles and then laugh. I see fathers greet them. Some are subdued; others are more expressive in their acknowledgment. The mothers stand in the rear. They then slowly, even cautiously, approach their daughters. The mother's eyes do not give away their emotions. I wish I could look behind their eyes and see what is inside them like I could clearly observe in their daughter's gaze of strength.

FRIDAYS ARE SUNDAYS, MORNING IS NIGHT

I was up before the alarm went off at 4:30 a.m., but this still counts as sleeping in. I've made a commitment to always sleep in on the weekends. The extra half-hour is precious and, whether it really makes a difference or not, I feel better rested for it.

It's Friday, "Juma," the Islamic holy day. The Ministry of Defense is closed, so the NATO advisors only work half-days.

Today is my Sunday.

The first hour of my day doesn't change, weekend or not. I have a roommate in a 8x15 foot cubicle, so I have to be careful not to wake him up. First, I reach for the head lamp next to my rack and switch it on. It lights my way to my locker, where I get soap and toothpaste, and then it guides me safely to the door. Once into the lit hallway, I turn off the headlamp and shuffle the remaining 30 feet to the bathrooms.

There are two bathrooms, each with two sinks, two toilets, and three showerheads. I'm up well before anyone else, so I have my pick of bathroom. Across the four toilets, there's only one toilet seat, and it sometimes rotates from latrine to latrine. I look in both bathrooms and choose the one with the toilet seat.

I'm convinced the scarcity of toilet seats is payback for all the times we forgot to put the seat down for our wives and daughters at home.

At the sink, I turn on my iPod and hit "Shuffle." I brush my teeth with a bottle of water and shave with the water from the tap (the water is disinfected for use, but you still shouldn't drink it). I usually only get two, sometimes three songs during my bathroom routine; it's the only time I listen to music, and I'm at the mercy of the "shuffle god" inside the iPod who (sometimes capriciously) chooses my early-morning entertainment. But I've made its job easy: I have no bad music on my iPod, only songs that give me happy thoughts. Today, I hit the jackpot with back-to-back tracks by Jack Johnson, the laid-back Hawaiian surfer with an acoustic guitar.

Johnson's song "Better Together" transports me to the 6th floor balcony of a condo in Indiana overlooking the Ohio River and the incredible skyline of Louisville. We had bought the condo because we spent so much time in Louisville, a city we had grown to love. Our house was at Fort Knox where I worked. Our home was the condo. I'm with my wife, Laura, my daughter, Madelyn, and a good but incredibly dumb dog named Gigi.

Gigi is a promise I made my daughter when I went to Iraq. She wanted a puppy and had been begging us for one for over a year. "Please, Daddy, I want a puppy," she would say with her enormous brown eyes. I agreed and said, "When I get back from Iraq, okay?" "Okay, Daddy," she said batting her eyes and doing her best to conjure a tear or two. Madelyn had learned very young how to master her father's heart. I went to Iraq and frankly forgot about the promise. When I returned, my mom and dad fulfilled my commitment and bought a dog. I came home thinking a large German Shepard or maybe a Labrador would meet me at the door, launch at me, and nearly knock me down with their size. That's not what happened. I came home to a four-pound, yapping white Maltese. To add insult to injury, they named the dog Gigi. I won't admit it to my friends, but I actually like her.

The three of us sat and looked out over the great river and heard the Belle of Louisville, a tourist steamboat that is docked at the harbor, blow its horn. We talked about what we were going to have for dinner and decided on pizza. Jack Johnson played softly in the background.

I can almost feel the sunshine. I'm there for one cherished moment. I towel off my face and look in the mirror, and I'm back in Afghanistan.

I dress in my Army running gear and walk down to the Goat Café to grab a cup of coffee. The Goat is open 24 hours, and I have my choice of "regular" or "strong" coffee. (I normally split the difference.) The sun is coming up, but the morning is still very cool; it almost smells like fall in the Midwest. It occurs to me I haven't seen a baseball score in two weeks, and I wonder how the St. Louis Cardinals are doing.

I take my coffee to my office. The room is empty and I sit at my desk and turn on my computer.

People ask for my BlackBerry number all the time, and they're amazed when I have to dig in my wallet and look at my business card to tell them. *Don't you have your own number memorized yet?* they ask. No. I do not. Laura taught me every memorization technique known to man so I'd remember her birthday; by now I have it committed to memory, but it only stuck after a couple of severe beatings. Madelyn stopped asking me for help with her math homework when she was in third grade. I am simply hopeless with numbers. With one exception.

Sitting there by the phone, I know without looking that 8094633376 gets me a dial tone and that 18002697241 connects me with AT&T. The sweet automated lady on the other end graciously thanks me for serving our country and then asks me to enter my ten-digit pin. After I do that, she tells me to dial the phone number I'm trying to reach and politely reminds me how many minutes I have left on my phone card. I don't even look at the phone pad as I dial the digits.

I lean back in my chair and take a sip of coffee while the line rings for the first time. I hold the receiver in one hand, and in the other I hold my cup of coffee. While the line rings I gaze at the screensaver on my computer: a picture of a tree in full bloom with every autumn color imaginable. It sits alone, set apart from rest of the woods, in one of my most beloved places.

Like the promise to get my daughter a puppy after my Iraq deployment, I told my wife that we would buy a house in Louisville for my Afghanistan tour. We had decided that we would retire in Louisville and that date was only a few years away. She had found a place, and I agreed to purchase it, even though I didn't know what it looked like. I had only one condition: it had to be as close as possible to Cherokee Park. Cherokee is a Fredrick Law Olmsted masterpiece, and a runner's paradise.

True to her word, Laura picked the perfect spot. A trailhead to the park is 40 feet from the backyard of the house; the picture of the tree was taken on one of her runs. She tells me it's *our* tree.

The phone rings three or four times, and in the time of those three or four rings, I am transported across 7,500 miles and eight and a half time zones, from Kabul, Afghanistan, to that place near the park (my new, forever home). Across three continents and an ocean, I am genie-blinked to a blissful reunion with a voice on the other side.

Who answers depends on who got the least time on the phone last. Today, Laura answers. In the background, our daughter calls her mom a "phone hog." In response, Laura reminds the 14-year-old of the 29 hours of labor she endured to bring her into the world. I'm completely enamored with their playful banter, and in a way they can't know now, I feel utterly loved.

I wish Laura happy birthday. She turns 31 (again) today. When one of my girls has a birthday, the celebration is about two months long. Laura's festivities begin on Halloween about the same time she starts bugging me to put up the Christmas tree, and the merriment doesn't fully end until after the new year.

Laura and I met a week after the birthday when she turned 31 the first time. We were married on Christmas Eve six weeks later. We experienced the full cycle of courtship in a single fall season.

She counsels me that her birthday is not until tomorrow. "Don't rush it," she says. "I'm aging fast enough as it is," she jokes. I remind her that, while it might be eight o'clock Friday there, it's four-thirty Saturday morning here. My wife, the eternal optimist, cheers: "Yay, two birthdays!"

As we talk, I click through a series of pictures Laura sent me. The photos show our tree with fewer and fewer leaves as the winter deepens. The last snapshot is of Madelyn next to the tree without any leaves left. The weather in the photo is overcast and cold, the winter gray. My wife and I prattle on about when the tree will start to bud in the spring and when it will be full and green again.

This is my favorite part of the day. Those moments are ours, just for us three.

I hang up and walk to the Warrior's Gym to start my "Sunday" morning run. Sundays are for longer distances, and today I plan to run the ¾-mile loop around Camp Eggers ten times. The loop is a lollipop: a long straight "stick" and then a side street that loops around and back. I've memorized every step of the loop but have yet to be bored.

I set off to run the "stick" of the lollipop, my favorite part of the course. I pass under 200 feet of green overhang—the only trees on the camp. For those 200 feet, I'm carried to other green spaces on other Sundays, like I might have been running through the green of Cherokee Park.

But, then, I'm back in Afghanistan.

THE LIST

It's been a rough couple of weeks.

I've allowed myself to get into a bit of a slump. The Cardinals' victory in the World Series was a short-lived joy amid inevitable melancholy—a natural part of the emotional ebb and flow of deployment.

To combat the ennui, I review notes I made in a black and white composition book on the plane to Afghanistan three months ago. On the first few pages of the journal I had scribbled lessons I'd learned from my previous combat tours. At the top of the page I wrote, "Remain optimistic and search for the good no matter how well it's hidden. Negativity is wasted motion." That aphorism was followed by: "Never forget, you volunteered for this."

A war veteran once said to me, "The realities of our lives are long periods of boredom, followed by brief but intense periods of excitement." I am one of the fortunate few who never bores, whether in combat or sitting alone in a room. There always seems to be something to occupy my mind. That has its advantages, but it can make it easy to get into a routine that rapidly becomes a rut.

At the end of a particularly slow day, I decide to organize my room—all 8x5 feet of it. During the clean-up, I find an Axe bath scrunchie my wife had sent me a few weeks back. When I'd gotten it in the mail, I'd looked at it the way a cave man would look at an iPod. I'd appreciated the thought but promptly placed it in the back of my locker. *No way I'm bathing with a man loofah*, I thought. *I don't even know how it works.*

I continue my cleaning and keep glancing over at the tidy pile of mesh. I feel like Tom Hanks in *Castaway* talking to his volleyball friend, "Wilson." "I'm not using you!" I growl at the pricey sponge from across the room. But I keep looking at it. Finally, I give up. I grab it, along with the other Axe products Laura had mailed to me, and go to shower. My buddy "Scrunchie" changes my life. I have never felt so clean and refreshed.

"Try something new, now and again; the variation will give you a new perspective, uplift your spirit, and break up the monotony."

I sometimes equate living on Camp Eggers to being in prison, minus the fear of being accosted by a muscle-bound predator who wants to call you Shirley. Aside from the obvious comparison—living in a confined two-man room within a walled compound—life on Eggers is deprived of other luxuries just by virtue of being in a war zone. To pick an example that's obvious when I'm stateside: I can't drive to the Bluegrass Brewing Company and enjoy a delicious Altbeir and Chef Mikki's "Wort Hog Wings." The menu at Eggers is severely more limited: broiled haddock, broiled chicken, baked haddock, broccoli, and pinto beans. It's not that bad, but let's just say it loses its luster fast.

I'd begun taking the "food is fuel" approach to dining, neglecting everything other than the bare essentials to nutritional survival. I sometimes ate peanut butter straight out of the jar and avoided the mess hall altogether. One day, a friend coaxed me to the "Goat" dining facility for dinner. I had my usual (haddock, of course). My buddy finished his meal and left the table. He came back with a cup of praline ice cream. My jaw dropped. Praline is my all-time favorite! I went and got a bowl with caramel and chocolate syrup and those sugary sprinkles. It wasn't crab cakes, but it sure did the trick.

"Savor the little things, no matter how small. They make a big difference when times get tough."

A couple of weeks ago I received the most gracious gift from my mother-in-law, Maureen. She sent me a homemade Halloween card. On the front is a picture of my daughter, Madelyn, at age three wearing a furry purple costume of Dot, the heroine from her favorite childhood movie, *A Bug's Life*. Madelyn is sitting on

Laura's lap wearing the costume, a flashlight in one hand and a plastic orange pumpkin in the other. She is squirming to be let go to begin the night's festivities.

There's a second snapshot: a younger version of me, hugging Madelyn, her chocolate-smudged mouth upturned in a mischievous grin. Later that night, she ran circles around the house until the sugar high wore off.

After reading the card a half dozen times, I propped it next to my computer. Mornings started and nights ended with Maddie in her Dot outfit. But then, thoughts of that Halloween led to thoughts of holidays that followed, especially the ones that I missed. I was always so happy to see the card, but I wanted so badly to be back, and the only way back was to wait.

After I got that card, my family began to possess my thoughts throughout my days—and mornings, and nights. I caught myself thinking about last year's Kentucky Derby when I should have been paying attention to my surroundings. Once, while leaving Camp Eggers, I caught myself thinking about Maddie's first high school dance—which I was going to miss—and I realized a few minutes later that I hadn't locked and loaded my pistol. That may sound minor, but an oversight like that is unacceptable for a professional soldier in theater; it's preparation for combat, which is your reality, and you cannot skip steps.

That was it. I come back to my office and read the list again from start to finish. I highlight the most important entry of them all.

"Your work may require you to lock out thoughts of the people you love. Your duty is to do your job and return to them."

I put the card inside my desk. I can open that drawer any time I need to.

FRIEND OF A BRAVE MAN

We stand out like American tourists in Afghanistan.

Our convoy of two up-armored SUVs skulks through Kabul traffic cautiously like a tortoise crossing a highway. Our driver, a young airman, cranes his neck to see as he maneuvers around Massoud Circle. He slows and swerves to avoid a woman in a blue burka crossing the street. She doesn't acknowledge the near-accident and floats by us like a phantom. Our Chevy Suburban's array of antennas and soldiers in full kit, sitting two each in the front and back seats, might as well be a flare on a moonless night. At every intersection I keep an eye out for motorcycles and Toyota Corollas, which are all over the country. They are also on the Be On the Look Out, or "BOLO list," as possible instruments for suicide attacks. I'm always alert, even when I'm with General Karimi, but on this trip I'm hypervigilant, like I just main-lined three cans of Red Bull.

It was different when I used to travel with Karimi, in his vehicle. When we approached intersections, one of the other four trucks in our fleet would block the traffic to let us pass, making a vehicle-borne suicide attack nearly impossible. If we had to stop, half the 20 Afghan commandoes who protect the general would get out of their Ford Rangers and disperse along the route to inspect cars and encourage them to start moving. The other members of the Chief's personal security detachment would stand as sentinels in the bed of the pickups and scan the horizon for any trouble ahead or behind.

I miss riding with General Karimi.

We had our best conversations during those 30- to 45-minute movements we routinely took to visit training sites in the Kabul area. On those days, before the "incidents," I would sit next to him without body armor—"slick," as he called it—and listen to him speak about whatever came to his mind. I most enjoyed his stories about the Kabul he grew up in, long before the wars, when the landscape was green and the city fresh and when Fridays were spent picnicking by the lake. "In those days, Afghanistan was *more* modern than it is today," he would say.

All his other reflections made that comment so ironic.

He'd talked about the Soviet occupation, U.S. support to the Jihad and the Mujahedeen, Russia's defeat and disintegration, the United States' subsequent withdrawal of support to Afghanistan, the spread of warlords, and the Taliban's rise. Mullah Omar consolidated power in Kandahar and Helmund and gave a home to Al Qaeda. In the time from 1979 to 2001—two decades—Kabul, along with the rest of Afghanistan, lost its modernity and all that was green.

In the time of the Taliban, Karimi was in exile in Pakistan. He was coordinating his family's asylum in the U.S. through the American Embassy, and he was scheduled to leave soon. Then, Osama Bin Laden attacked the U.S.—an operation that was planned within the safety of his sanctuary in southern Afghanistan. Omar enacted the ancient Pashtunwali code and refused to give up Bin Laden. The U.S., sanctioned by NATO and propelled by the will of the American people, counterattacked.

The Chief once told me, reciting an Afghan proverb: "If you can't be a brave man, be the friend of a brave man." There is nothing more important than the support of the international community in helping create stability, he would say. There is no scarcity of Afghan bravery for fighting, but for democracy to work in Afghanistan, Karimi believes, other nations have to be courageous enough to commit.

The world's devotion to the mission in Afghanistan, and the devotion of the U.S. in particular, has been declining for some time. During an early-morning call home recently, I had to settle down my daughter, who was frustrated at a friend for saying that the war was over since the U.S. was out of Iraq. Madelyn had to remind the college freshman that soldiers were still in Afghanistan. Then her friend said back, "Isn't it the same thing?"

I wish it were just the girl's youth speaking for her, but I've heard folks my own age say that.

The recent turn of events has shaken our resolve more than ever. Marines were filmed urinating on the corpses of insurgents and burning Qur'ans, and there was a spike in green-on-blue attacks, where Afghan soldiers turned on their Coalition partners, killing or wounding them in alleged retribution for the sacrilege. The final straw, it seems, was an attack where two Ministry of Interior Advisors were shot and killed at close range by an Afghan co-worker. The day after the shooting, with scattered reports in the media, I received a cryptic email from my former War College Professor, Chuck Allen, that read simply: "Situation report, please." It was code for *are you okay?* Though I'm an advisor in the Ministry of Defense, the news didn't clearly spell out where the two dead officers actually worked.

I can't ride with General Karimi anymore because of new protection measures in response to those attacks. I'm now required to sit next to another Coalition soldier when traveling in the city or elsewhere. This means I often have to move in Allied convoys to locations where the Chief is going. When I can hitch a ride with the Afghans, I have to move with a battle buddy, or "Guardian Angel" as the second soldier is called. We are all fully kitted and in body armor. There's not enough room in the Chief's vehicle so we squeeze into another truck.

Karimi's security personnel, who have protected me for the last eight months, sometimes point at my protective gear and rib me in good humor, asking, "Don't you trust us anymore?" We laugh, and I try hard to hide my embarrassment.

I glance out the window of our SUV and flinch as a Corolla nearly clips our rearview mirror. When the enhanced security order came out, I told my supervisor, "My relationship with the Afghans *is* my body armor. I feel less safe wearing the body armor now than I did without it." I never felt more secure than when surrounded by General Karimi's guards. I am nearly always by the Chief's side; if someone was intent on killing me, they'd be hard-pressed to break through that circle and do it.

Our convoy passes through the secure gates of North Kabul International Airport, the location of the ISAF Joint Command headquarters and our destination for the evening. I get out of the SUV and take off my gear. I see General Karimi's entourage and walk over and shake hands with the Chief's son Zia. Zia serves as Karimi's chief of security, but he is leaving soon to attend the Infantry Captain's Career Course at Fort Benning, Georgia. There is a strapping, intense Afghan soldier next to him. I recognize the young officer from the General's home security detachment.

Zia introduces us. "This is Captain Sher Hassan," Zia says. "He will be the chief of security when I leave."

We shake hands and, without another word, Sher Hassan moves off with purpose to check on his soldiers. Zia and I walk quickly to catch up with the Chief, and Zia tells me, "Sher Hassan has fought in almost every province in Afghanistan. Once, a sniper up in a building had his platoon pinned down. He left his men with only his pistol and entered the building alone. He fought his way to the top floor and shot the gunman once in the eye."

Zia pauses. "Sher Hassan is a very brave man," he says, adding nothing more.

COMING BACK

I think of the Vietnam vet in the airport when I deployed to Afghanistan who asked me, "Just getting back?" At the time, I shook his hand and just said, "Going." I want to see him again and shake his hand and have him ask me that same question, because this time I'm coming back. My tour is done.

I want to go to a bar with the old man with the cap full of ribbons, drink beers, and compare war stories like I did with combat veterans after my other deployments. In between sips and our tales of adventure, there would be pauses where we would say nothing and just stare off into the distance. It is in that silent space where we would be most connected.

I've spent two of the last six years deployed. Compared to some, I've had a break. There are others who have yet to deploy. For whatever reason, the odds didn't work out, and they will have to pause uncomfortably when asked where they served during America's longest war. I feel deeply sorry for them that they missed out on this great adventure.

Two days ago, the Taliban's Spring Offensive was lit like a sparkler on the Fourth of July. It fizzled brightly for a short moment, and just as quickly it faded. Shakespeare's Macbeth would describe the insurgents in their rush as "…a poor player that struts and frets his hour upon the stage and then is heard no more: it is a tale told by an idiot, full of sound and fury, signifying nothing."

I could not have been more pleased before redeploying that I was afforded the opportunity to once again experience what Winston Churchill said best: "There is nothing more exhilarating than to be shot at without result." The hapless insurgents fired 20 feet above our compound, missing all of us by a wide margin. Still, the snap of a passing bullet creates a "fight or flight" response and, no matter how close, a perfect clarity of focus. I was having a grand time, but not everyone was.

I may never forget the sound of my wife's voice when I called the evening of the attack to let her know I was all right. I married a strong, independent, and highly successful woman, but the tone of the person on the phone was unfamiliar. Although I knew I had to read for clues, I couldn't tell exactly what was wrong. In a moment, it became abundantly clear. She began to softly sob and then broke into a full cry.

The news channel she'd been watching had finally done her in. I think it's a safe bet that the media never specified, as part of the coverage she watched, that no civilians were killed in the Taliban's attack and that only a handful of Afghan policeman died. No Americans were dead or seriously wounded. The Coalition didn't even participate in the fight. But everyone at home whom I love dearly had no idea what was really going on.

In the end, it is Laura and Madelyn who have selflessly served. I was simply afforded the opportunity to do my work the way I was trained. That is the great irony: as soldiers, we are often given an overwhelming amount of credit for doing our jobs. However, those who *truly* sacrifice are the ones at home who only vaguely comprehend what we do and whose only source of information is what's on television.

I feel an overwhelming sense of guilt. Sometimes soldiers are remarkably selfish in their selflessness. We take great, almost obsessive, pride in our values of loyalty, duty, respect, honor, selfless service, and personal courage. We have to. It is what sets us apart from other occupations. Soldiering is a profession, after all. Our nation cannot survive without men and women willing to do warrior's work. However, adherence to the values and their application in war can take on an intoxicating quality. Coupled with the thrill of combat and the sense of belonging to something bigger than oneself, there is no drug more addictive than war.

My guilt is that I will miss it.

I pack my bags and wonder: *When will I be back?*

LIVING ON AMBER

When I turned 51, I decided to revive a tradition that I'd inconsistently celebrated over the last several years. The tradition required that I do something uncommonly difficult on my birthday. By "uncommonly" the activity had to be physically demanding, requiring mental toughness to see it through to completion. The intent was to start the year off on a positive note by accomplishing a challenging task.

I briefly considered using a colonoscopy I was scheduled to have the week prior to my birthday as such an event. The preparation phase that occurs the night before the procedure certainly met the criteria. I had to drink four quarts—eight total pounds—of an electrolyte mix to prompt the cleansing process. Gulping down the nasty, chalk-tasting liquid was a grueling feat. I felt as if I swallowed a basketball and thought, "This is what it's like to be pregnant and in labor." I was reminded of my wife as she gave birth to our daughter. Sitting next to Laura in the hospital, I went to gently stroke her arm, and her head turned almost 360 degrees. Wild eyed, like Linda Blair in *The Exorcist*, she snarled with a deep and raspy voice, "Touch me again, and I will kill you." I had similar feelings as I lay on the couch praying for God to end it all.

Ultimately the colonoscopy was ruled out as my birthday challenge, since it was something I had to do, forced on me by cruel Army policy that requires men 50 years of age to have one done. To truly meet the spirit of the tradition this had to be a self-imposed hardship. So, I chose to move 51 kilometers by foot—a "klick" for every year of my age. Fifty-one kilometers is roughly 32 miles, and I had not run more than 14 miles in a while. This would not be easy, but I had run that far and many more miles in years past. I thought, *if I can run 108 miles like I did at the Pac Rim 24-hour race, this will be a piece of cake.*

Location is an important aspect for these kinds of events. There must be some stimulation to help guard against the weariness constant movement can incur. In this case, the route is an easy decision. I choose to run Cherokee Park's scenic 2.43-mile loop, which is right next to my house. This route will give me the ability to enjoy the park's beauty from the outside looking in. It is January, and the cold has defoliated the trees, giving a fresh and unique perspective of Cherokee's wooded interior. With the leaves gone, I can see tributaries to creeks I had no idea existed. Rock formations jut from hillsides like pre-historic sculptures. Wildlife cannot hide, and creatures of all types scurry about looking for food. Winter, this season typically thought barren, delivers a whole new world for me to enjoy. I have to complete 13 laps to achieve the distance. For some the monotony would be unbearable, but I'll be thoroughly entertained.

Henry David Thoreau wrote in his book *Walden*, "I went to the woods because I wished to live deliberately, to front only the essential facts of life, and see if I could not learn what it had to teach, and not, when I came to die, discover that I had not lived." I choose to run Cherokee Park for the very same reason.

I get up early the morning of my birthday and pack a plastic bag with Gatorade, gels, and PowerBars to sustain me during my run. I walk to the start point, but I'm really not feeling it. My birthday, the 10th of January, concludes over two months of celebration in my family. The festivities start on November 5 with my wife's birthday and run through Veteran's Day, Thanksgiving, our anniversary, Christmas, and New Year's Day. But, this year has been different. I had redeployed from Afghanistan in late April, and the party had actually started then.

Since coming back, I'd drunk and ate what I wanted. I'd lived as an Epicurean and refused myself nothing. Deployment denies many pleasures. Upon return, you are starved for them, but it comes with a price. I'd gained 10 pounds and felt soft, not only around the midsection; my willpower had become fluffy, as well. Small things, like the cold and rain, that didn't bother me while deployed, made me whimper. Inconveniences such as extraordinarily long traffic lights triggered hissy fits. I was becoming the person I despise most: a whiner.

As I stretch, bracing myself against one of the stone bridges that span Cherokee's Beargrass Creek, I begin compiling the list of excuses why running 32 miles without training is a bad idea. The best reason I come up with is that I could get hurt. (During that 24-hour race I tore my meniscus, which required surgery.) Other justifications range from being tired, feeling a cold coming on, and thinking that doing this was an absurd notion in the first place. Finally, I make the decision to just start running and see how I feel as time progresses. I zero out my watch. Then, with my first step, I press start, and the journey begins.

I divide the loop into six parts characterized by the topography of the route: two hills, two associated downhill sections, and two flat portions. The two hills are significant. Combined, they represent a climb of nearly three-quarters of a mile. The ascent is gradual, but late in the run the scramble up the inclines can seem like a hike up Mount Everest. As a result, those hills will become a focal point of the event. I will further compartmentalize the effort by counting up to seven laps and then counting down the last six. This is similar to how I got through deployments that are typically 12 months long. The count up to the halfway point is always the hardest. However, placing waypoints in the year, like my anniversary, holidays, birthdays, and other special occasions can make it easier.

Sometimes that technique works; other times, not so much.

I chose my wedding anniversary of December 24 for my waypoint for my last two deployments. For my 10th anniversary, I was on a forward operating base in east Baghdad. After a day of patrolling, I got back to the base and made a call home to my wife. We were on the phone for a couple minutes making plans for my mid-tour leave that was approaching. I was standing outside near a bunker when a familiar sound distracted me from our conversation. I said, "Oh shit," dropped the phone, and dove into the dugout. Several mortars had hit not far from my location. When I retrieved the phone, the connection had gone silent. It took a while to get back to my wife. When I did, I told her I was alright. There wasn't much to talk about after that.

We had our 15th anniversary while I was in Afghanistan, the same day a coalition member was killed by an Afghan police officer during an "insider" attack called a green-on-blue. By the time I phoned home, the news had reached the media, but the victim was described only as an "advisor" to Afghan forces. I was serving as an advisor to the Afghans. That was another call home that didn't go so well. My anniversary is not a good waypoint for me. If I deploy again, I'll try something else.

The first part of the route is flat. The even surface allows me to find my rhythm and breath. As I pass the dog park and rugby field, it occurs to me that I haven't read one single book since returning home. I laugh to myself, recalling Dean Wormer's chastisement of the fraternity boys in *Animal House*: "Fat and stupid is no way to go through life." I approach Golf Course Hill and think, "I really need to find something to read." The first part of the incline is steep. I struggle a bit, feeling like Flounder at the hand of Neidermeyer's malice. I add that I need to drop some weight, as well.

The two hills in Cherokee are monsters. On the climbs, I concentrate on staying loose as the lactic acid pools in my thighs. "Keep the knees high," I say to myself, dropping my hands to my side to shake them out. I learned this from high school cross country. My coach, Rod White, would tell us, "You have to learn to relax when you're in pain." There is a similar saying when rehearsing close combat marksmanship, "Slow is smooth, smooth is fast." The tendency is to get jammed up in the frenzy of battle. When you're in a firefight, the mind works like a summer storm. As lightning strikes in the horizon, clouds roll in while thunder rumbles in the distance. There comes a sprinkle and then the heavens open and rain falls in buckets. In the next instant, the sun appears burning away the clouds to bestow a pristine sky. What you do in the first few moments when engaged with the enemy is the difference between finding shelter or getting caught in the downpour. One must be in a careful hurry in those instances when there are a myriad of decisions that have to be made at once. A poor or ill-timed choice can be catastrophic—like the time I nearly shot Akmed in Baqubah before our mission to get food in Baghdad.

After the hills come the descents. Running long distances requires patience so you don't go out too fast. Pace should remain steady. It's easy to get caught up and carried away on the downhills. That has always been my challenge. I first earned the nickname "Frantic Fred" for my reckless approach to downhill trail running. The moniker stuck, but for other reasons. Of the four cardinal virtues, I have always struggled with temperance. Wisdom is a very close second. However, most of my unwise decisions were a result of a lack of restraint. It never occurred to me that being in a careful hurry applies when safely home.

I threatened to punch a fellow colonel in the throat once. I knew I was wrong immediately after I said it. So did the other colonel who reported me to my boss. I received a letter of concern for my unprofessional actions. The next step would be something much more harsh that could ultimately end my career if I didn't do something to change my behavior. My supervisor mentioned PTSD when he talked to me. I wish that was my excuse. The issues I have with anger have absolutely nothing to do with my war experience.

My wife, a licensed psychologist, would disagree. She has repeatedly told me that I changed after my deployments. I listen to her politely, but I believe I inherited the malady like the color of my hair and skin tone. Johnsons are, by nature, hot tempered. I'm not crazy and I don't have PTSD. I just get upset when someone, particularly if they are my rank, does something unethical. I finally gave in to my wife's pleas to get help. She told me, "It's fine if you don't believe you have PTSD, but you have to do something before you throw your career away." I scheduled an appointment with a psychologist, figuring that would get everyone off my back.

Enya hummed in low tones from speakers hidden somewhere in the waiting room, and the mystical Celtic rhythm was calming. I scanned from wall to wall, corner to corner, and across the ceiling looking for the source of the sound. I couldn't locate it. This was playing out just as I expected. The meditation music, the room's soft colors, the framed pictures of pastoral bliss all made me want to take a nap. I thought, "This is going to be great.

The door opened, and a slight man of medium height stepped in. "Colonel Johnson?" he asked. I was wearing khakis and a polo shirt to disguise that I was in the military. My cover was blown. He shook my hand firmly and said, "I'm Larry. Please follow me."

Dr. Larry Raskins certainly looked like a shrink with his close-cropped gray beard and light blue button-down shirt. The experience thus far pretty much confirmed every belief I had about therapy. I had resisted it for years. First, I didn't think I needed it. Second, I believed if I had a problem, I could fix it on my own. Finally,

I thought it was quackery. When my wife got her Ph.D. in psychology, I quickly softened my position on the last point, but I still pushed back against her every time she mentioned that I should see someone.

We walked into his office. He directed me to a chair. He took a seat straight across from me. An empty coffee table separated us. Dr. Raskins crossed his legs, placed a notepad in his lap, looked at me and said directly, "What can I help you with?" That's a strange question, I thought. It took me off guard so I answered honestly, "I have issues with anger." I told him. "So, you lack self-control. I thought soldiers were disciplined," he replied immediately. I wanted to slug the guy. I thought we were going to talk about my childhood, bedwetting, and if I had an unnatural love for my mother. I was prepared for all those questions, having rehearsed the answers on the drive over. "Where's the couch," I thought, adjusting myself in the chair. I felt like I was sinking into the leather cushion. Raskins said nothing. He sat there like a little gray bearded sphinx. His posture was rigid. His eyes had a fierceness about them that contradicted my initial impression of the man. The silence became very uncomfortable and I finally said, "When something upsets me I explode sometimes. I guess you're right. I can't control it."

His response was as rapid as his first: "Yes, you can. Everything that pertains to your mind is within your control. You can't control what goes on outside your brain, but you can manage your thoughts and how you respond to situations that cause you to be angry." He placed his notepad on the table, then leaned slightly forward and said, "The first thing you must learn is to stop, think, and arrest your thoughts before you act. Think about a prudent response to the emotion." And so it began.

Raskins gave me tools to address those times when I was nearing the tipping point. To my surprise they really worked. Sometimes I would improvise on what he taught me though. For example, with the colonel I threatened to punch in the throat, every time he spoke, as much as he aggravated me, I pictured him as a gnome in a conical hat. When he said anything, I just smiled. My sessions with Dr. Raskins lasted nearly six months. Initially I saw him once a week and then every other week. Then, I stopped all together. I was fixed, I thought. Several months later I deployed to Afghanistan.

I long for the flat segments of the course. I reach eight laps, nearing the mystical "wall" of 20 miles. I start to walk for short periods of time, especially on the hills, which have become increasingly more difficult. For me, walking requires less concentration. My mind wanders. I start obsessing about food. I am starving and have run out of PowerBars. It feels as if I'm in a Looney Tunes cartoon where one character envisions another character turning into some delicious meal. Everything is becoming a cheeseburger to me. There is a manhole cover on the road not far from Hogan's Fountain Hill. When cars drive on it, they make a very loud noise. As I'm about to take the first bite of my imaginary snack, a vehicle rolls over the sewage drain. It makes a booming sound, and I jump off the road into the ditch. I laugh at my silliness, but I am reminded that I need to pay closer attention to what I'm doing. It's later in the day now and traffic has picked up.

Soldiers are taught three stages of alertness when in a combat zone. Green is when there is no danger. Contact with the enemy is very unlikely. The situation is normal, like a leisurely drive in the country. Amber is when a threat is possible but not expected. It is the state of mind most people have when they approach a busy intersection and look both ways. Red is the highest state of preparedness when an attack is imminent. This is likened to a person's mental state during an impending head-on vehicle collision. One can stay "Red" for only so long before you burnout. That level of hypervigilance is exhausting. However, you can maintain the "Amber" category almost indefinitely. This color code also applies to a soldier's weapon status: Green, the magazine is removed; Amber, a magazine is loaded in the weapon, but there is no round in the chamber; Red is locked and loaded.

On my base in Afghanistan there was great debate on what weapon's status service members should maintain. The policy was "Green." There had been a string of "negligent discharges" where someone accidently fired their weapon. Usually, this occurred during the process of clearing it before entering the secured compound. It's called "negligent" because there should be no accidents when handling a firearm. The only time it should let off is when that is the intention of the person with the gun, unless it is the result of a mechanical error, which is rare. The reason people have negligent discharges is because their state of alertness is "Green" when their weapon is "Red." This should never happen.

My weapon was always "Red" in Afghanistan, and I tried, as much as possible, to keep my state of alertness on "Amber." A fellow officer reprimanded me once because he saw that my magazine was in my 9mm. I was, in fact, in violation of policy. He was correct in his rebuke of my weapon posture. I politely told him to go screw himself. At a nearby base, an Afghan Air Force officer walked into a room where nine NATO members were having a meeting. They were all "Green." The Afghan was "Red." He shot and killed all nine of the personnel. Al Qaeda and the Taliban are crafty foes who had infiltrated the Afghan National Army. This asymmetrical strategy had resulted in nearly as many coalition deaths by insider attacks as occurred on the battlefield. The reality was the combat zone was everywhere, even on the confines of our bases. The downshift from "Amber" to "Green" should be a conscious decision, one made deliberately. However, the Sirens of complacency can lure you into a "Green" stupor, if you're not careful.

I deployed to Bosnia in 1996 as a part of a NATO force to help stop the Serbian genocide of Bosnian Muslims. We would prevent the slaughter by putting ourselves in between the Serbs and those they intended to exterminate. In the unit I was assigned to, not a single shot was fired with the intent to kill. This was a totally different kind of war than any I had previously experienced. To be totally honest, the biggest danger was the boredom. One time a buddy who was a Croatian interpreter and I were walking down a road in the winter. Snow was on the ground. I had to go to the bathroom so I took off to run in a field and do my business behind a tree. As I was running my buddy yelled, "Fred, No stop!" I said, "What dude? I'm trying to take a whiz. He then said, "That sign you passed it said 'minefield' in Serbian." Anyone who thinks a man can't stop peeing midstream has never pissed in a minefield. I will tell you that stupidity caused by monotony will kill you as quick as a bullet. I nearly died in Bosnia because I was on "Green." That was a mistake I would not make again.

I have been on "Green" for most of my run. I need to shift to "Amber," not only to be watchful for oncoming cars, but also to be alert for signals my body is sending me about bad running posture, tenseness of the hands and neck, and abnormal deviations in my stride. "Red" will come later when thoughts of quitting creep into my mind.

I have been moving for over four hours. I have reached the marathon mark, 26.2 miles, and it starts to rain. I'm out of food and fluids. The temperature is dropping. I stop my watch and head home to get wet weather gear along with some Gatorade to rehydrate. I think, "A marathon is not a bad day's work. Five more miles in this rain is going to be rough." The rain picks up as I shuffle to the house.

I reach my back door just as the rain really starts to come down. I slip off my soaked gloves. My fingers feel arthritic from the cold. I have trouble guiding the key into the lock. I cuss. I feel the anger swell in its usually starting place, my stomach. I say to myself, "Slow down, dude. It's just a door lock." I take a breath as the key finds its way inside the deadbolt.

I come into my house through the entryway of our home gym on the bottom floor. We have a treadmill, elliptical, spinning bike, dumbbells, and mats. The fitness equipment is positioned so we can look out a large window that faces the woods. On those rare days when the weather prohibits exercising outdoors, we can vicariously enjoy nature by looking from the inside out like fish in a bowl. I start to warm up a bit as the rain splatters against the glass pane. The droplets obscure the view as I turned my attention to getting myself squared away so I can finish the run. I had put a change of clothes on a counter in the fitness room just in case. The weatherman had predicted showers, but for later in the day. I thought I would have been done before it got bad.

Just as I'm pulling off a sock, my left calf muscle seizes up. The knot causes my foot to flex upward. The right leg does the same almost causing me to fall backward. I let myself down on to the mat, pushing both feet forward to try and release the spasm. I massage both legs. I curse the cramps as anger swelled. "This is great. Shit! Two laps left and...." I take a deep breath and soften my inner voice. "Slow down. The cramps will go away. You need some electrolytes. It's all good. Chill," I say to myself as the water from the rain mixes with my sweat and pools on the floor around me. After working the muscles with both hands, I grab a Gatorade, twist the lid off, and take a long gulp; I remind myself that I made the choice to run. "Anger is wasted motion," I whisper. I shift from Amber to Red. The enemy has surrounded me. I flip my mental selector switch from safe to automatic and I destroy every reason for doubt that entered my mind.

Satisfied that they are all dead and I will finish this damn run, I go back to Amber.

I drink an entire quart of Gatorade and eat a PowerBar. The cramps subside. I change my clothes, socks, and shoes. The rain has lightened up, but I put on a Gore-Tex jacket just in case. I walk down the hill from our house. When I hit the asphalt on the road to the park, I start to jog. The short rest has tightened up my muscles. I feel a little stiff, but not terribly bad. When I reach the bridge that marks the beginning of the route, I start my watch. Two laps left. On the first loop I walk the hills, but I finish the last running. My Ironman Timex reads six hours, a quarter of the day spent in Cherokee up and down those hills. Coming down Hogan's Fountain Hill and into the flat finish one final time I pick up speed and think of the old man in Longview, Washington, at the Pac Rim 24-hour run. "This is a timed race," he said. "So is life," I thought, "We just don't know when the clock stops." I get to the end of my birthday adventure and continue to jog home. I open the back door. I step in and go to "Green."

FOUR WARS

Everyone in my office has gone home for the day. Our building is empty except for the security guards downstairs and a few other people who work the nightshift.

I'm alone in my office, in the dark. I sit back in my chair and stare at the email I've been writing and re-writing for the last hour. I can't decide whether or not to send it.

The email is addressed to Major General Steve Townsend, the commanding officer of the 10th Mountain Division. I had served with General Townsend twice before, once as his executive officer (XO) when he was a battalion commander and the last time in Iraq as his deputy when he was a brigade commander. The subject of the email: "Assignment."

It is 2013 and I have been back from Afghanistan for about a year, working in an administrative assignment with the United States Army Recruiting Command. I have just learned that the 10th Mountain is deploying to Afghanistan. I want to go with them. The email I've written is a request for General Townsend to help get me reassigned to his division so I can deploy again.

I have not talked to my wife about it, but I am sure she won't be surprised when I tell her.

My first months back from Afghanistan had been wonderful. I returned in time to go to the Kentucky Derby, that festival of horses and bourbon that had first brought us to Louisville. We had bought a beautiful house near Cherokee Park, paradise to runners—and therefore to me as well. My daughter was happy in her school and felt that she finally had a real home after moving eight times in 12 years.

Life was good.

Maybe it was too good. In the few months after I returned, evenings and weekends were filled with laughter, and eventually became more subdued. Sitting on the couch at home, there would be long silences, and my wife would ask, "What are you thinking about?" Afraid to tell the truth, I would just say *work*. Sometimes, I'd get impatient with her questions, then angry. In those moments, I felt something come to a boil and build up inside of me, and more and more often the steam would break through the lid and scorch everything nearby.

In my office, I slowly lean forward, held my finger there a moment, and press *Send*.

If Townsend can work it out, this will be my fifth time going to war. I hope that he can. Now truly alone in my office, I wonder why I miss it so much.

Then I remember.

★ ★ ★ ★ ★

I loosened the straps of my pack, slipped it off, and let it fall at the base of a big tree. I turned around and started to sit, as though the pack were a chair, but halfway down, my aching back muscles and tight hamstrings gave out and I just plopped to the ground. Vinnie O'Neil came over and did the same. Vinnie took out his canteen, unscrewed the top, and took a long drink, water trickling from the side of his mouth. He gasped for air again, then put the cap back on the canteen and placed it between his legs. He stared straight ahead into nothing, his breathing deep and tired. He looked at me as he sighed. *Brutal*, he said.

Vinnie took off his thick, black Army-issue glasses. Sweat channeled down his freckled Irish nose, around his mouth and down past his chin. He tried to wipe the glasses on his shirt, but it was no use—his whole uniform was completely drenched. He put the glasses back on, still wet and grimy, and sighed again.

Just brutal.

It was the fourth day of our 100-mile foot march. We were in the 2nd Battalion, 22nd Infantry Regiment, which was a part of the 10th Mountain Division. We were light infantry, which meant we would walk a lot. This was our training—trudging up and down the dirt roads and trails of the Adirondack Mountains of upstate New York in August. The first day we marched 30 miles; the next two days were 20 miles each. When Vinnie and I sat down, we were halfway through the day's 18-mile march. The next day would be the last of the exercise, and the shortest day with a cool 12-mile march back to the garrison at Fort Drum.

Vinnie lived across from me in the Bachelor Officer's Quarters, or "the Q" as we called it. We were the last two men standing; all the other lieutenants had either moved out to apartments or gotten married. Vinnie and I spent nights after work, drinking Bud and sipping Scotch, either watching war movies or talking about the latest book we had read. Sometimes we even recited poetry. (I sometimes joked that the poetry had been the reason the other junior officers left.)

Vinnie and I signed into the battalion about the same time in the summer of 1986. I was the anti-tank platoon leader and he led the mortar platoon. Vinnie believed that fate had somehow landed him the position, since his father was a mortar man in World War II. He kept a picture of his dad in uniform on his dresser.

Vinnie was from Boston. He was a proud Irishman, a good Catholic, and he could down more beer in an hour than most of his comrades could drink in a week. But he was no drunk—and the only sure sign Vinnie was buzzed was that his wit came out like a switchblade. If you pissed him off, he'd cut you down with four or five words.

Looking at him, you'd never think he was an infantryman. He was short and stocky. Vinnie had to take twice as many steps to keep up. I always thought of him as Barney Rubble in the Flintstone Mobile, with Barney's feet churning in place until they caught traction and the vehicle took off. Short legs or not, Vinnie was always there at the end. He might be bent over, puking his guts out and cussing, but there was no quitting in him.

Vinnie broke his stare into nothingness, took the canteen from between his legs, and held it in front of him as Hamlet held Yorick's skull. He admired the canteen for a moment, then quoted Kipling in his best penny accent: "You may talk o' gin and beer when you're quartered safe out 'ere, An' you're sent a penny-fights an' Aldershot it: but when it comes to slaughter, you will do your work on water, and lick the bloomin' boots of 'im that's got it." Vinnie smiled at his own timeliness, brought the canteen to his mouth again, and drained it.

We'd been sitting by this tree for a few minutes when we saw a figure in the distance walking up the trail. It was our chemical officer. From nearly the moment we'd started the foot march, he began to fall behind. Every time we stopped, he would catch up; when we got moving, he immediately fell behind again. He shuffled, barely lifting his feet off the ground. Every couple of steps, his foot would catch a root and he'd stumble and nearly fall. His Kevlar helmet sat on the back of his head (were it not for the chinstrap, it would have fallen off miles ago). His face was chalk white except for his red-blotched cheeks and hazy dark eyes. The lieutenant's mouth was barely open, and his panted groans were partly audible. He shuffled passed us, not saying a word, and disappeared further into the wood line.

I was about to make a joke about the *Night of the Walking Dead* when we heard someone call for a medic. It wasn't a cry for help, but more like someone yelling over a crowd to a bartender. It was a demanding and familiar voice; Vinnie and I got up and ran over to it. A couple dozen soldiers had encircled our battalion commander, and we edged our way in to see. The commander was standing, still wearing his load-bearing equipment and Kevlar, but he was unbuckling his trousers.

Being summoned, the medic pushed his way through the bystanders, then stopped so fast he nearly tripped over himself when he saw the commander. The medic put down his first aid pack and said, "What's wrong, sir?"

The commander replied, "Doc, I got a bad problem."

The medic asked, "What's wrong? How can I help?"

"You got some moleskin, son?"

The medic responded that he did have moleskin—an adhesive medical tape typically used to prevent foot blisters.

"Gimme a whole shit-load of it," the commander said. He grabbed his crotch and said, "My cock and balls are chafed to hell. It's killing me."

The medic, now holding the tape, looked at the commander dumbfounded, as if to say *and what exactly do you want me to do about it?*

The commander dropped trou, took the moleskin from the medic's hand, and unrolled an arm's length of the tape, stretching it out as he pulled, bare ass and jewels hanging freely with everyone around. He took his knife and cut the tape, then wrapped his penis and testicles snugly, pulled up his pants, and handed the moleskin back to the medic. Then he ran in place to test it: knees high, pumping his arms up and down like he was sprinting. "Much better," he said.

This battalion commander was an older lieutenant colonel; he had a round, pock-marked face, black hair, and even blacker eyes. He was medium height and he looked almost like Sitting Bull. When he wore his physical training uniform, only his Popeye-sized forearms distinguished him as an athlete. His legs were scrawny, particularly his right leg (his calf had been blown off during one of his Vietnam tours), but even with one calf instead of two, he could run us all into the ground. He first served in the Dominican intervention in 1965 as a junior enlisted man, then he did a tour in Vietnam as a sergeant, then another one or two as an officer. His last combat tour included the Battle for Hunger Hill.

Rumor had it that he'd been a guinea pig for a top-secret military drug-testing program when he was a private. They said he was given LSD or something and that was why he only slept two hours a night. The guy loved to train. During live-fire exercises, rounds would hit so close but the old man would start laughing, and he'd scream at the top of his lungs: "Bring it closer! Get it closer, you gun-bunny motherfuckers!"

We would follow this guy through the gates of Hell.

Our commander said, "Okay, let's ruck up and move out." We moved out laughing every step for the first half mile. No one ever considered quitting a foot march, except of course the chemical officer; he'd be put on a vehicle later that day, too weary to finish. He would ride the rest of the way while we walked. I don't remember ever seeing him again. That was best for all of us, but especially that young officer who could never un-quit. He had no place in our ranks.

The next day, after our short 12 miles, we marched through the gates of Fort Drum onto a field where we did physical training. The battalion commander led the way, walking bowlegged by now. He climbed atop a wooden platform in the middle of those grounds, and he gave the order to fall out. Five hundred soldiers gathered around him.

The commander was not an eloquent speaker by any means. He spoke the way a boxer would jab: sometimes the words missed, but when they made contact, his follow-up punch might knock you out.

He began by congratulating us on completing the foot march. "Not that you had a goddamned choice," he added. A bunch of us laughed.

He stepped forward on the platform so that the toes of his jungle boots jutted just over the edge of the platform. He leaned forward like he might dive headfirst into the crowd. He scanned the formation, seeming to look at each one of us, and we stared back at him in unison. Then he said: "You've probably been wondering why in the hell did we just walked 100 miles. I'm going to tell you why."

He stood erect and grabbed his crotch. (That was more a habit than a gesture; it seemed the old man always grabbed at his balls.) Then he said, "Raise your hand if you ever thought of quitting." Not a single hand went up.

"Yeah. That's what I thought, he said. "Why do you think that is?"

He paused for a second. Then he said, "It's because you didn't want anyone to think you were a pussy." The entire formation burst out laughing in unison. He waited for the laughter to die down, released his hand from his testicles, and finished what he had to say. "We walked a hundred miles to learn that one lesson. You're not a pussy. The men who finished the march with you are not pussies. Because pussies don't win wars."

The commander released the enlisted men to their supervisors to clean their weapons and return their equipment. He told the officers to stand by. As the soldiers marched off to their company areas, the commander issued his instructions to the officers. "Clean your shit, make sure your boys are good to go. Beer call at the club at 1600."

Drinking alcohol was not really optional in our battalion—at all. Officers were expected to drink beer at a minimum, if not also hard liquor (they preferred tequila and whiskey). Wine was barred from the Officer's Club except for officers' wives. But the drinking culture of the 10th Mountain is probably best described by the drink the commander invented for us—and which was, in fact, a requirement of initiation for every officer in the battalion.

To make a "Mountain Mule," take a beer mug and start with a generous pour of tequila. Then, fill it the rest of the way with Tabasco. You have to finish your Mule in one chug or it doesn't count—and you don't receive the coveted battalion coin or, worse, the acceptance of the rest of the team.

At 1615, I was on my second beer and leaning against the bar at the O Club, as we sometimes called it. Vinnie and I were hanging with a new second lieutenant who had just been assigned to the battalion, who had just completed the 100-miler with us. He made the march without any problems, so he was okay by us; he just had to throw down a Mountain Mule.

Vinnie asked the newbie if he had ever heard of our Division Commander, Major General Bill Carpenter. The lieutenant said *no.*

Vinnie loved telling this story. I had heard it a million times, but it got better each time, especially if Vinnie had downed a Mule or two.

"So we're sitting at a table over there, having a few beers," Vinnie says, gesturing to another part of the room. "All of a sudden, the hair on the back of my neck stands up, like they say it does right before lightning strikes.

"I turn around and there's Napalm Bill Carpenter, standing right next to us. We jumped to attention so fast that we almost knocked over our chairs. But he told us to take our seats, so we did, and he sat down with us."

Vinnie took a sip of his beer and asked the new officer if he had ever seen *Platoon*. The newbie said *of course* (there wasn't an infantryman alive who hadn't seen it at least six times).

"Well, you know at the end, where the commander calls in an airstrike on his own company because it was getting overrun?" The lieutenant nods.

Vinnie says, "That story is based on how General Carpenter got his Distinguished Service Cross in Vietnam. He killed some of his soldiers doing that, but he saved most of their lives. The guy is a no-shit hero."

The newbie sips his beer. He seems mildly impressed, and he says *cool*. Vinnie shakes his head and waves a finger in front of the lieutenant's face and says, "Oh no, no. I'm not finished." Tipsy or not, Vinnie is a bit disgusted with this guy's failure to appreciate the setup. Vinnie takes a long drink of his beer and prepares the closer.

"You see, when Napalm Bill sits down with us, he lights up a Camel. No filter. He doesn't say anything, but he takes a deep, long pull on his cigarette and blows the smoke right in the face of our signal officer—our SIGO—and the SIGO coughs and fans his hand in front of his face. Carpenter looks at the SIGO and says, 'Does my smoke bother you, Captain?'

"The SIGO says, 'Why yes, sir, it does bother me.'

"So Carpenter takes another pull on the cigarette—damn near inhales the whole thing down to the end—then blows a huge cloud of smoke right back in the SIGO's face and says *then stop breathing.*"

I chuckled because I knew what was coming. Vinnie said, "The SIGO didn't know what to do, so he took a deep breath and he held it. He looked like a blowfish. His face turned blue, and he nearly passed out. When he finally let it out and started breathing again, Napalm Bill gave him a look like he was going to rip his head off. So the SIGO literally *ran* out of the O Club."

We laughed like the SIGO was some kind of patsy, but the truth is that we'd have done the exact same thing. We were scared shitless of General Carpenter—and nearly all of the leaders in the division, for that matter. They were all war heroes.

Our division operations officer was the most decorated officer still on active duty; our commander had a couple Bronze Stars for Valor and something like four Purple Hearts; the division chief of staff was a Special Forces officer who supposedly killed an entire squad of North Vietnamese without a gun. The list just went on and on.

They all had Combat Infantryman Badges—CIBs, as they called them. I often caught myself staring at that badge on the uniform of anyone who'd earned it. How you got it was pretty straightforward: you had to be in an infantry unit engaged in combat, and you had to be in close proximity to a hostile confrontation with enemy combatants. The Combat Infantryman Badge was simply designed—a rifle with a wreath around it—but it was a Holy Grail to all of us. We pined desperately for the chance to earn our CIBs and assume our place among these giants.

Someone put money in the jukebox, apparently, because Bon Jovi's "You Give Love a Bad Name" began to play. As it did, Vinnie nearly snorted beer out his nose. He grabbed the new lieutenant by the shoulder and said, "Oh, you've got to hear this one."

Vinnie had an endless catalogue of stories. He loved it when we got new guys so he could tell them all over again. Vinnie says, "We had a kid get shot during training."

The newbie's eyes get big like *what the fuck*, but Vinnie tells him, "Nah, don't worry about that. Brigade commander says it's not good training unless someone gets shot every now and then. And this has only happened once, as far as I know, and it was this one soldier in Alpha Company I'm telling you about."

"Well, anyway, the kid gets hit, but he's all right. Doesn't die or anything, and he's back in A Company, but every time the soldiers in his company see him, they sing, *shot in the chest and you're to blame, you give training a bad name,*" Vinnie sings, repeating the altered lyrics. Vinnie and I laugh; the newbie doesn't. He starts looking for an excuse to go somewhere else, but about that time, someone pulled the plug on the jukebox anyway.

The gap of silence was immediately filled with the unforgettable voice of the battalion's most senior enlisted man, Command Sergeant Major Klaus Madsen. "*Awwright*, heroes, fill up your beers. *Da* commander is *ah* coming." Madsen was a Dane whose accent had an odd sense of poetry to it. The long, drawn-out vowels and the *da* sounds that intermittently replaced the letter *T* had the rhythm of a train on a track. Madsen had moved from Denmark to Canada in the early 1960s; he came to the U.S. to go to Vietnam (while loads of U.S. citizens were skedaddling to Canada to *avoid* the war).

He thought that part was hilarious. He'd say, "Aww *mudderfucker*. I would see dose sorry *mudderfuckers* going north when I was *ah* going south and said what sorry *mudderfuckers* dey were, dat dey wouldn't fight for America. Udder than *da* crappy beer it's *da* best goddamn place in the world."

The commander and Madsen would drink until one, the other, or both were comatose—and that was long after every other officer in the battalion had puked, pissed his pants, and passed out.

The commander walked in and yelled over to the bartender—a much older, but still beautiful and big-bosomed redhead we each secretly dreamed about. "Get me a beer and two Mules," he shouted. With perfected precision, the barkeep had the beer and two Mountain Mules on the bar before the commander could step over to pick them up. He took a sip of the beer, put it back on the bar, picked up the two Mountain Mules, and called the new lieutenant up.

The newbie angled his way through the crowd and took his place next to the commander. The commander handed him a Mule and put his hand on the lieutenant's shoulder. The commander said, "Great job this week, men. You deserve to let loose a bit. Work hard, play harder. But first, we have some business to take care of."

The commander tightened his grip on the newbie's shoulder and looked at him. "He passed the first test on the foot march, but here's where the rubber meets the road. Bottom's up." Lieutenant and commander brought the mug to their mouths at the same time, tilted their heads back, and drank until their glasses were empty. Other than their eyes reddening and tearing up from the Tabasco, neither one showed any sign of distress. They slammed their mugs on the bar to the cheers and applause of the rest of the room.

The commander dug in his pocket and brought out a battalion coin. He placed it in the lieutenant's hand with a firm handshake and said, "Welcome to the team."

The commander then grabbed his beer, raised it above his head, and shouted our battalion motto: *Lead with Courage!* We repeated the words, *Lead with Courage!* and we drank.

The drinking went on long into the night. As the alcohol flowed, the commander and Command Sergeant Major Madsen began telling war stories. We gathered around them like kids around a campfire. Their tales of bravery, adventure, and loss mesmerized us like flames and legends. We could smell the battlefield like the smoke of a fire pit. When the night was over and there were no more stories to tell, we left wondering when we'd be able to tell our own war stories.

But our battalion never went to war. The closest we came was a month-long deployment to Honduras. We were only there for training—but the Sandinistas and Contras were still fighting in neighboring Nicaragua. There was instability in El Salvador and Guatemala, given the drug trade that was beginning to prosper.

It was for those reasons, I assume, that we were each given a single 20-round magazine of live ammunition for our M-16s. We had to keep it in the breast pocket of our uniform, and not loaded in the weapon. We each called that 20-round clip the "Barney Fife magazine." There was never any reason to use it—but we had been issued live ammunition in a potentially dangerous environment, where war *could* erupt. That possibility alone was almost consoling to us.

After about three years, we were all forced to leave the battalion, to leave Fort Drum. Soon we'd be promoted to captains and head to Fort Benning, Georgia for more senior-level infantry training, after which we'd be assigned to other units around the Army.

Fort Drum had been a tough place, but it made winners. Vinnie got out of the military as a captain and became an award-winning author of science fiction, horror, and mystery novels. Many officers from the battalion were eventually promoted to colonel, to command their own battalions and brigades, and to receive valorous commendations for their parts in the wars then to come. Our commander was be promoted to major general. Two of these officers became high-level advisors and policy experts at the Pentagon. One major who joined the battalion near the end of our tour would later become a four-star general and the most senior-ranking black officer in the Army.

One of our fellow lieutenants, Chad Buehring, said to me as we left, "It's never going to be this good again. There's no way it can be." Whether he's right or wrong, it's hard to say; he was killed in Iraq during the Second Gulf War.

* * * * *

I was en route to my next assignment at Fort Campbell, Kentucky in December 1989. Right before Christmas, the U.S. invaded Panama for Operation Just Cause. Those of us who didn't go to Panama felt we'd missed the big one. We cried into our beers and lamented our outrageous misfortune. Later, when those officers and soldiers who participated in Just Cause came back with CIBs, we suffered miserably—cursed to be less than men (and certainly less than those warriors).

My next unit was the 101st Airborne Division (Air Assault) at Fort Campbell. The Airborne Division maintained its historical legacy, along with all of the bravado. The 10th Mountain was a plough horse; the 101st was a thoroughbred. Which description was a compliment and which was an affront was for each person to decide.

I was assigned the 3rd Battalion, 187th Infantry Regiment: the Rakkasans. The first person that welcomed incoming officers into the battalion was the executive officer. Having been tormented by the XO back at Fort Drum, I had some serious apprehension going into the meeting.

While sitting in the waiting room, I heard shouting coming from the XO's office. *Shit! Fuck! Damn it to hell!*

A man who looked 60 opened the door and stepped into the waiting room. My first thought was that he might be a janitor. He held a wrapped Burger King cheeseburger in one hand, and with the other he was trying to wipe mustard from his brown Army-issue T-shirt (which wasn't tucked in in the back). His attempt to wipe away the stain only smeared it more. He looked at me, turned back to the office, and waved me in with his cheeseburger hand.

He sat down with a flat *thud* and a sigh. I stood at attention, as is military custom until one is told to stand at ease. Such an order was never given.

"All right, all right, let's get this over with. Welcome to the battalion, *yada yada yada*," he said, waving his hand. "You're going to be the Battalion S1. Do you know what that is?"

I was startled at the question, so I didn't answer right away. He sighed again and said, "That's the personnel and administration officer for the battalion. It's called an S1. Got it?" I finally nodded and said *yes, sir.*

"Okay, that's all. You have an appointment to meet with the battalion commander next. Shut the door on your way out," he said. I could see, as I turned to walk out, that he'd already returned his attention to the cheeseburger.

My next stop was the reigning adjutant's office to check in, then wait for my appointment with the commander. The departing S1 was a young Mississippian, already balding, but with an impeccably starched uniform. He too was eating Burger King, but he had the fish-filet sandwich, and unlike the XO, he used a clean napkin to neatly wipe his mouth. He stood slowly, slightly bowing, and extended his hand after wiping it with the napkin. It was a very formal gesture of welcome, but his palms were moist and his grip weak.

"So, my relief has finally arrived," he said, his voice more nasal than I was expecting. "And not a minute too soon. The commander is in a mood and fit to be tied. I'm afraid your meeting with him may not be pleasant. It seems his wife's flowers did not arrive at the scheduled time, so I've been the object of his ire."

The departing S1 sat and gestured widely, offering me a seat with a royal wave of his hand. "I see from your records that you were a platoon leader for 30 months but have only six months on staff as an assistant operations officer," he said. "To be frank, I don't think you have the necessary experience to be an adjutant. I'm sure you'll learn quickly, though. *Sink or swim* as they say."

I think I smiled. I didn't say anything.

The S1 continued. "And, fortunately, you have an excellent NCO who will run personnel and administration while you see to—" He cleared his throat and handed me a sheet of paper. "—the commander's needs."

The page listed all of his former duties, which were now my duties. They included keeping the commander's refrigerator filled with Fresca, seeing that the commander's vehicle was always clean, writing thank-you letters, and picking up the commander's laundry at the cleaners. It was a long list. At the very bottom—highlighted in bold capital letters—was written:

"MAKE SURE THE COMMANDER'S WIFE RECEIVES FLOWERS EVERY MONDAY MORNING BY 10:00 A.M."

The S1's phone rang, and he quickly picked it up. He nodded and then said, "Yes, sir. Right away, sir." He hung up the phone, looked at me, and said, "I'm afraid the commander cannot see you today. He is not feeling well, and he is leaving early for the day." I looked at my watch. It was 1:15 p.m.

"You may continue your in-processing, and I will set up an appointment for another day."

When I finally met the commander, I felt a vicious coldness inside, the kind of feeling you'd have if you were caught out in the December rain with no jacket. The sharp chill would fade when I wasn't around him, but it came back every time I heard his voice. There could not have been anyone more my opposite at Fort Drum. He was a very small man with delicate features. He was perfectly kept, with a starched uniform and fingernails I swore had to be manicured. His hair was much longer than I'd ever seen on an officer; I had a short high-and-tight. He was much too young to have been in Vietnam, so he didn't have a CIB. He never talked about war, his vocation; instead he focused on social engagements, on finding opportunities to meet with the Commanding General and other senior officers.

It wasn't long into the job before I realized that Chad was right and nothing would ever compare to Fort Drum and the people with whom I'd served there. One day I went to the garrison lawyer, who helped me draft a letter of resignation. I signed the letter and kept it in my office drawer, waiting until the right moment.

I was done. I had to get out before it went too far and I did something stupid, like beating the shit out of the greasy XO (or forgetting flowers for the commander's wife). I finally submitted the letter to my battalion and brigade commanders, and they endorsed it. But before it was sent forward to the Division headquarters for final processing, Iraq invaded Kuwait and we were alerted to deploy. I told the commanders that I changed my mind and really didn't want to resign, so they returned the letter to me. I tore it up and went to the First Gulf War with the Screaming Eagles of the 101st.

The First Gulf War had two distinct phases: Desert Shield was the build-up to combat operations; Desert Storm was the actual war. During Desert Shield, we defended Saudi Arabia and prepared for our eventual attack against the Iraqi Army to drive them out of Kuwait.

I'd never experienced heat like Saudi Arabia. It was like a blow-dryer was always turned on HIGH and aimed at you, except this blow-dryer also coated your body in sand. In theater, a shower was (more or less) a bucket that hung over your head and, at the pull of a string, dumped water that was either scorching hot or freezing cold, depending on the time of the day. As soon as you toweled yourself halfway dry, the sand would stick to your body like plaster and grit. Never once in that deployment did I feel fully clean. But like other adverse conditions, the sand and sweat had a bonding effect for the soldiers and my fellow officers.

It began to feel, at least sometimes, like I was back at Fort Drum again.

Fortunately, I saw the commander only rarely. Not long after we arrived in country, he contracted what we called the *NVDs*—nausea, vomiting, and diarrhea. The NVDs were common, but most soldiers recovered in a couple days; the commander had it for about three months. He spent the majority of Desert Shield in the aid tent hooked up to IVs.

One day, the battalion command sergeant major and I were driving around, checking on the troops, when we saw the commander's driver tip-toeing to the laundry. He was holding a set of desert camouflage fatigues still wet with shit, his fingertips and arms extended fully away from his nose. We pulled up beside the driver and the CSM asked him, "What in the hell are you doing, Sergeant?"

The young NCO turned his head to the Sergeant Major, keeping his nose up and away from the uniform, and said, "The commander shit himself again, so he told me to wash it."

"Oh, I swear to...."

The CSM shoved the shifter into park, sprang out of the car, and snatched the uniform from the Sergeant. With the same hand he threw open the tent's door, stepped inside, and marched over to the commander, who lay groaning on a gurney. He threw the soiled fatigues at the commander and said, "Sir, the next time you shit yourself, you call me to wash your uniform. But *never* order that soldier to do it again."

After a month-long air campaign against Iraq, the ground war was ready to begin. The day before the attack, I noticed the commander wasn't on the manifest to accompany the main attack force by helicopter. I asked another captain—a staff officer who'd been in the battalion much longer than me— if there was some mistake. He shook his head in disgust.

"I just heard this from a guy at division headquarters—so it may or may not be true—but, evidently, our fearless leader had a premonition that he'd die," the captain said. "So he's made the courageous decision to drive up with the logistics convoy and, um, *follow* us into battle. You know, to ensure our *much-needed* supplies arrive on time."

I was not surprised. In fact, it confirmed everything I'd believed about the commander from the moment I'd met him. There'd been another rumor that he was scared enough to sleep with a loaded .45 in his sleeping bag—which, according to another buddy, went off by accident one night. The round blew out the bottom of his sleeping bag and went through the tent. And the only reason the stray bullet didn't hit something (or someone) else is that the commander had ordered the engineers to build up a berm around his sleeping area to protect him from indirect fire.

Evidently, the commander didn't see what I saw. On February 24, 1991, I was 29 years old—and I walked out onto the Saudi Arabian desert to a stretch of attack helicopters as far as the eye could see. I rode a Black Hawk helicopter 155 miles into Iraq, across the Euphrates River valley where the most ancient civilizations rose and fell.

This was the war that I had trained for. It was everything that I hoped a war might be. Desert Storm was the second-largest invasion in U.S. history; only Normandy was greater. It was a war the world knew, a war that might nobly liberate a country from its tyrant, a war that took me halfway around the world to a land I'd otherwise never have seen.

On the day of the invasion, we landed 35 miles east of Babylon. Even then, the irony was not lost on me—that we had come to the cradle of civilization to wage war on one.

The one thing that was missing from the experience . . . the kind of combat our commander and CSM Madsen (the Dane) had described when I was at Fort Drum. There were some units that fought difficult battles, but not the unit I was assigned.

The only engagement was on the second night. I was checking our perimeter when, in an instant, an entire company of infantryman opened fire on a suspected enemy convoy. The firing lasted only a couple minutes, then the company commander sent out a platoon to assess the damage. When they returned they made their report: one truck full of onions had been destroyed, but no sign of any occupants. Signs in Arabic down the road warned any innocent civilian traffic not to travel this way; the vehicle had failed to yield the warning.

The next day I was walking to the command post and saw a strange shape behind the only vegetation around. I walked over behind it and found a dead Iraqi. He was dressed in a white *thawb*, or "man-dress" as we called it. He had a single bullet wound in his stomach; the blood had wrapped around his midsection like a wide belt. His eyes and mouth were open and flies were beginning to swarm around his face. I wanted badly to brush them away, but I couldn't.

★ ★ ★ ★ ★

Desert Storm lasted 100 hours. It was over before I knew it. I came away thinking that war was easy—maybe that it was fun, even.

I would later regret that thought.

Some have said Desert Storm wasn't a real war. I can only assume they think so because it wasn't long enough or violent enough, or that not enough people died. Yet for those same reasons, I would argue that it was a near-perfect war.

The strategy of the war was born out of the Powell Doctrine, where the American military uses overwhelming forces and strike capabilities (with the widespread support of the American people) to achieve a vital national security objective—and one that has a well-planned exit strategy in place.

We came home to neighborhoods and city streets decorated with yellow ribbons. On the Fourth of July I marched, proudly displaying my newly sewn-on CIB, with my company in a ticker-tape parade in Atlanta, Georgia, in celebration of our victory. Small children tugged at my uniform waving miniature flags. It was the first time someone ever said to me *thank you for your service.*

I remembered Chad Buehring, my fellow lieutenant at Fort Drum who said it would *never get any better than this.*

I thought *yes, Chad, it can.*

* * * * *

Not long after we returned from Iraq, I assumed command of Bravo Company. About a month later we got a new battalion commander. His name was Lieutenant Colonel David H. Petraeus.

In our first meeting, I briefed Petraeus on Bravo Company with an overview of key personnel, our readiness status, and planned training events. Briefings like these were very important—not only to give that commander an accurate assessment of the unit's capabilities, but also, to connect with that commander and gain his trust.

During the meeting, I wasn't sure if I was making the impression I'd hoped. Petraeus always seemed to be a slide ahead of me in the briefing. He would say *got it* before I'd even finished what I was explaining. I felt like I was briefing a supercomputer. When I was done, Petraeus pushed the briefing packet to the side of the table and said, "All great stuff, Fred. Just great."

I thought, *shit, I really blew it.*

He walked over to his desk and picked up a book. He brought the book over and handed it to me as he sat down. It was a hardback with yellow pages, some folded at the corners, like you might find at a garage sale. "Fred, have you ever read Richard Haliburton's *Royal Road to Romance?*"

I said *no sir.*

He told me, "Look it over and tell me what you think. Great meeting, company commander. Well done." And the briefing was over.

I left thinking: *what the fuck?*

About a week after briefing Petraeus, the phone rang at six-thirty on a Saturday morning. I had been up for a while getting ready to work out. I answered the phone. "Hello?"

"Fred, this is LTC Petraeus," the voice said. "Meet me at the battalion headquarters. Let's go for a run. See you in 15." And then he hung up.

When I arrived to the battalion area, Petraeus was waiting, doing pushups. He greeted me as I got out of my car and asked if I knew a good 10-mile run route. I said I did, and he took off running, without any stretch or warmup.

We started at a reasonable pace, the kind that makes running conversation easy. Petraeus was an efficiently-built animal, something you knew would survive from its apparent strength and swiftness. There was no wasted space on his body; his form was perfect as we ran. I noticed right away that his arms moved straight forward and back, assisting him forward like pistons, but without the up-and-down you see in inexperienced runners. Rather than clench his fists and tense his body, he kept his hands relaxed like he was cupping a robin's egg as he ran. His head didn't move, even when he spoke, and he seemed to breathe fully without opening his lips. His eyes stayed focused ahead, never down.

"So what did you think of Haliburton's book? What was your major take-away?" he asked me as we started an uphill climb.

Richard Haliburton was a major author of travel and adventure books in the 1930s. He swam the full length of the Panama Canal, circumnavigated the globe in a biplane, and eventually died while attempting to sail a Chinese junker from Hong Kong to San Francisco. Haliburton's *The Royal Road to Romance* was one of many books he wrote about his adventures.

"I think Haliburton was an adrenaline junkie," I decided to say between breaths. "He went from one adventure to the next to get his fix. He's a great writer and his stories are awesome, but I'm not sure what he accomplished other than living it. It's not, you know, like he discovered the cure to cancer or anything."

My answer caused Petraeus' stride to get a bit longer. I adjusted mine back up. He asked, "Did you like the book, Fred?" I nodded *yes* as Petraeus moved the speed up yet again.

"But *why* did you like it, Fred?"

I was starting to feel the stress of the run now; we were running six-minute miles. I tried to relax despite the fact that I was getting my ass kicked. Petraeus' questions became more and more abstract, requiring me to give more details in my answers. I tried to make succinct replies between breaths, but he wouldn't accept most answers until I'd expanded on them.

With about a mile left in the run the questions about the book finally stopped. We were running a sub six-minute pace now. It had become a race—never mind that Petraeus outranked me and was my boss. At a street corner nearing the finish line, I cut him off and took the inside, closest to the turn; in the process I bumped his arm. At the next turn, he gave me an elbow, throwing off my rhythm. I could taste copper in my mouth from exertion and fatigue; I could barely breathe and my muscles were screaming for me to stop, but the battalion headquarters was only a couple hundred yards away. Those last agonizing seconds were a dead sprint—and as we crossed the imaginary finish line, where we'd started the run about an hour previously, Petraeus took a final long step and leaned forward like he was breaking the tape.

The battalion took on this new character under Petraeus. However congenial and cooperative he demanded commanders be, they could be sure that nearly every training event would incorporate some kind of competition. Like the 10th Mountain Division, we worked hard and played hard—but our "play" was distinct from Fort Drum, where we drank tequila and Tabasco against madmen until we leaked. Under Petraeus, by contrast, the officers routinely had dinner parties he called the "Renaissance Rakkasan"— where the officers cooked their best gourmet food and their wives sat on the judging panel.

Petraeus introduced an intellectual quality to warfare—and even the preparation for combat—that I had never considered. Sometimes his thoughts were far beyond our immediate comprehension, but he taught us how to think rather than what to think. His only requirements for the way we trained our soldiers: make it realistic and make it challenging.

Petraeus' first autumn in command, Bravo Company's soldiers were running through a combat training exercise with live ammunition. One afternoon, Petraeus was there to observe, so I escorted him and a brigadier general who was our assistant division commander. The three of us followed one squad (some distance behind) as they tossed a hand grenade into a bunker, cleared it with fire, then moved to another bunker to repeat the drill.

I watched as the last soldier cleared the bunker. He was trotting back to his squad for the next assault, rifle in hand, when he tripped and fell. When he hit the ground, rifle still in hand, it went off.

A moment later Petraeus grabbed at his chest, stumbled forward, went down to his knees, and fell cold onto his back.

The General took a quick step over, glanced at the wound, and said, "You're gonna be all right, Dave."

"Fred, he's been shot," the General said, stepping over Petraeus. "I'm going to my helicopter to call in a Medevac."

I sprang over to Petraeus and unbuttoned the blouse of his uniform. As I did, a small trickle of blood ran from the entry wound, a tiny hole in the front of his chest. I gingerly rolled him to his side to look for an exit wound and he groaned; there was a gaping hole in his back as big around as a coffee mug. He had a sucking chest wound; when I laid him flat again, he started spitting up blood and pink tissue.

Two of my soldiers, Specialists Smith and Curtain, rushed over and shoved me aside. "We got this, sir," Curtain said. They sealed the wound with bandages and plastic bags so Petraeus could breathe, then tied off the bags with knots. He was placed on a stretcher about the time the Medevac arrived, and he was taken to the helicopter was he was transported to Vanderbilt University Medical Center for surgery.

After the Medevac departed, the general called me and the officer in charge of the exercise over to his helicopter that was starting up to take the him back to the division headquarters where he would have to give a report on the incident and check on Petraeus' status. The general had been my brigade commander at Fort Drum. He was the officer Vinnie O'Neil told the story about to the newbie at the O Club. The one who said *it's not good training unless someone gets shot every now and then.* The irony was not lost on me as I stood at attention in front of him and awaited his guidance.

The general was about six feet five and towered over both of us. He had commanded a company in Vietnam and received the Silver Star. His piercing blue eyes looked straight into your soul in a mesmeric way that demanded your immediate respect and obedience. I had, on more than one occasion, stood in the way of that gaze at Fort Drum and was just as possessed by it on the range at Fort Campbell that day as I was as a young lieutenant. The general said to us, "I want you to do an after-action review with the soldiers and leaders about what happened. Have the squad go through a couple blank fire iterations and then I want them to do the exercise again with live ammunition." The officer-in-charge of the range was dumbfounded. He had served in a mechanized unit in Germany prior to coming to the 101st, and this was unheard of in most Army organizations. He thought the general had called us over to relieve of us of our commands. He incredulously asked him if he meant the soldier that shot Petraeus, as well. The general looked at him with his blue eyes and said, "Especially that soldier." He then got on his helicopter and flew away.

Petraeus recovered from his wound and returned to the battalion less than a month later. Evidently, his doctor told him that he was not ready to leave, so Petraeus did 50 pushups and said *it was time he got back to the battalion*. He called me into his office before physical training the morning he returned. I reported to him, and he told me to come in and take a seat. I was fully prepared to be formally reprimanded. That is what normally happens in serious incidents such as what occurred on the range.

Petraeus was reviewing some papers and then put them down. He looked at me for a moment and said, "You know, Fred, there is a rumor going around the battalion. The rumor is that the soldier was actually aiming at you and hit me." I thought he was serious for a second and then he laughed, "That would speak volumes for both your bad leadership and your company's poor marksmanship, wouldn't it? Loosen up, Fred. That's a joke." He laughed again and told me, "Mistakes happen. Get back to work, company commander."

★ ★ ★ ★ ★

For my next assignment, I became a coach and trainer at the Joint Readiness Training Center at Fort Polk, Louisiana. My job was to mentor platoon leaders during two-week simulated combat scenarios.

This was a highly selective assignment; only successful company commanders were chosen. Virtually every captain that served in this position was later promoted to major and nominated to receive advanced education. Sure enough, I was selected for promotion to major (before most of my peers) and I was given the opportunity to attend the Army's Command and General Staff College at Fort Leavenworth, Kansas.

Then I fucked up.

While still at Fort Polk, I was arrested for driving while intoxicated. I received a General Officer Memorandum of Reprimand, or a GOMOR. I was taken off the promotion list for major and expelled from the Command and General Staff College. Later in the year, a military tribunal would determine my suitability for continued service—and whether I'd stay in the Army.

My career was over. No one survives a GOMOR.

The shame I had caused myself and the disgrace I had brought to my team was too much to bear. So I volunteered to go to the Balkans with the 1st Armored Division, which was stationed in Germany.

I was working at CALL (the Center for Army Lessons Learned) at Fort Leavenworth then. There was an officer in my unit who was scheduled to deploy for the upcoming mission, but his wife was pregnant. I asked him if I could go in his place; that way he could be there when his kid was born. He reluctantly agreed.

Right before I got on the plane for Germany, he took me to the side and said, "I really appreciate what you're doing. Really, I do. But I have to know something."

I shrugged. "What is it?"

He put his hand on my shoulder. He said, "Brother, are you going to Bosnia because you hope you'll be killed?"

I looked him right in the eyes and said, "Are you kidding me? No way."

The difference between guilt and shame is that guilt means *I made a mistake*, but shame means *I am a mistake*. I didn't feel guilty about what I'd done; I felt ashamed of myself. There was a part of me that believed I'd already died. I believed that I would be out of the Army by summer and that the only way I could salvage any dignity was to deploy again—to join the ticker-tape parade one more time.

In Germany, I was assigned to the 1st Brigade, 1st Armored Division. I was an *attachment*, which meant that I was not officially assigned to the brigade. I came on right before the deployment with barely any time to assimilate into a seasoned group of officers and NCOs that comprised the "Ready First Combat Team" (the

Brigade's nickname). My job was to capture lessons learned from the operations there and write them as articles for CALL.

The Brigade Commander, Colonel Greg Fontenot, accepted me into Ready First with open arms, but he made his one condition clear. "You can do all that lesson-learned writing stuff on your own time—at night, when everyone is asleep. During the other 20 hours of the day, you will be one of my plan's officers. Go see the operations officer. He'll get you set up with the other planners."

I would not be an outsider looking in. I would be a member of his team. It was as simple as that.

I spent nearly a month in Germany preparing to deploy. I was assigned in the plans section to help the senior planner, Mark Vera, another captain. The office where we worked was aptly called the dungeon. There were no windows; it smelled of old, wet magazines and stale books.

Mark told me, "You'll get used to the smell, but after a while you'll start to feel like a mole. The sunlight is a killer when you come out from down here, so make sure you get some sunglasses." Then, Mark brought me swiftly up to speed on the operation we were undertaking.

Mark reminded me a lot of Vinnie O'Neil. He was sharp and quick-witted, and yet he looked nothing like an Army officer.

The 1ˢᵗ Armored Division, and "Ready First," was part of a large NATO force of 60,000 troops. This would be the first NATO out-of-area deployment since NATO's creation in 1949, and it would be the first time the U.S. had allied with the Russians since World War II.

Our collective mission was to implement the mandate of the Dayton Accords. The Serbians had murdered thousands of Bosnian Muslims and were fighting Croatians for control of Bosnia-Herzegovina. Bosnia is where WWI had started with the assassination of King Ferdinand. During the overview of the mission, called Operation Joint Endeavor, Mark asked me if I had ever heard of the grunge rock group Offspring. I said yeah, I liked them.

Mark went over to a tape player on his desk. He said, "I hope you really, *really* like 'em because, when we get stumped developing the plan, this is coming on." He pressed play and the lyrics from an Offspring song played over and over again: *You gotta keep them separated. You gotta keep them separated. You gotta keep them separated.*

Mark turned off the tape player and explained to me that the plan called for the dispersing of the warring factions (the Serbs, the Croatians, and the Bosnians) to their designated areas so we could implement the other stipulations of the peace treaty.

Mark turned the tape player back on. *You gotta keep them separated. You gotta keep them separated. You gotta keep them separated.*

He jumped up and down on his toes miming jump-rope, swinging his arms wildly and bellowing at the top of his lungs, *you gotta keep them separated, you gotta keep them separated*, you *gotta keep them separated.*

I thought, *the dungeon has really messed this dude up.*

Mark turned the tape player back off. He put his hands on his knees, panting, then pulled up his shirt to wipe sweat from his face. "Damn, I have to get in shape. *Whew.*"

Still catching his breath, Mark asked, "You're a light infantryman, right? Airborne, Ranger, blah, blah, blah, probably never been in a tank before?"

Almost insulted, I just said *yeah, why?*

Mark said, "You've got to put all that shit behind you. None of that stealthy, sneak-up-from-behind, cloak-and-dagger stuff here. The only thing the Serbs understand is brute force."

Mark pointed to the only picture on the dungeon wall. It was a painting of a formation of M1 Abram Tanks and Bradley Fighting Vehicles in the attack during Desert Storm. Apache helicopters hovered overhead firing missiles at the enemy below. Artillery shells exploded in the distance. Iraqi armored vehicles smoldered in fire and black smoke. It was a picture, if you will, of U.S. military might.

Mark stepped over to the painting and waved his hand over it in reverent display, as if he were beckoning me to inhale the destruction. "The politicians are calling this a "peace operation," whatever the fuck that means. It's bullshit. We have to go to Bosnia prepared to fight and kill every one of the bastards—and if we're not ready for that, the Serbs won't take us seriously."

I had one of my most memorable Christmas Eves in Heidelberg, Germany, a week before deploying. I went to a candlelit midnight Mass at the most beautiful cathedral, where the light from dozens of candles seemed as stars in a dark and stony night. They played Bach's *Jesu, Ode to Man's Desiring*, and I decided then (and put into writing later) that that song would be played at my funeral if I died in Bosnia.

I spent Christmas Day with a German family I had met. They invited a beautiful woman about my age to the festivities. We hit it off well and had an incredible night together. But the holiday was over too soon, and I had to return to my unit to deploy.

We'd put the heavy equipment on rails and positioned it in neighboring Hungary and Croatia. The remainder of the soldiers would depart Germany by way of Austria, then Hungary, and finally our destination of Zupanja, Croatia, on the Sava River and along the border of Bosnia-Herzegovina.

In Desert Storm, I participated in the longest and largest air assault in military history at the time. We flew a brigade's worth of soldiers, seats out over the barren desert for 155 miles, from Saudi Arabia into the Euphrates River Valley. But during Operation Joint Endeavor, I was going to war on a bus.

The day we left Germany, I woke up with an itchy crotch. I was poised to take the crabs to war, too, but for some handy Lindane shampoo I got from the physician's assistant. I got on the bus, took a seat by a window, and nodded off thinking *what a gal.*

I woke up several hours later as we crossed from Austria into Hungary and then stopped the buses. We'd had to store our ammunition in the lower cargo compartments of the bus until we were out of Austria, since they had become pacifists and didn't support what we were doing in Bosnia. Everyone filed out of the bus, retrieved their ammo, stepped back on, and rolled into Zupanja at sunset.

It was the 29th of December, 1995. The Sava River had swollen from record-setting rain and snow, so our crossing was delayed. That night, I slept with a couple other captains in a small tent.

New Year's Eve, I woke to the sound of gunfire. I grabbed my weapon and ran out into the cold. Mark Vera was standing calmly a few feet away, sipping coffee out of a canteen cup. He was wearing Gore-Tex pants, a black stocking cap, and the liner to his field jacket, plus black gloves. Mark slurped the instant Folgers from the cup quickly so he wouldn't burn his lips and tongue. Between sips, he was looking up at the sky, where green tracers from AK-47 fire shot across the horizon like falling stars.

"Celebratory fire," Mark said before taking a longer slurp. "They know they have one more night of doing what they want. Stupid bastards don't consider that the rounds have to fall to earth somewhere." He took another slurp.

"Happy New Year, brother," he said, still staring into the dark.

"Happy New Year," I said, watching the gunfire light up the sky.

The next day, we crossed the Sava.

Colonel Fontenot would be one of the first to meet with the Serb commanders in our area of operations. Mark and I went with him to help set up for the conference. The meeting was set for 11 o'clock in the morning, but we arrived about three hours early.

The main conference room was several tents all connected together. In the middle, there was one long table with chairs, and chairs were lining the walls of the tent for other observers. Tanks and Humvees with .50-caliber machine guns surrounded the briefing area. Crews manned their armored carriers in full battle gear. The official reason for their presence was the protection of the meeting's attendees, but that didn't make it the real reason.

Fontenot looked around and nodded his head approvingly. "You guys did a good job," he said to us. "You better have or you would have gotten the bat."

We all chuckled. This was a reference to Fontenot's old Louisville Slugger that hung on the wall above his seat. If someone said something stupid in a staff meeting, Fontenot would stand up, grab the bat, and assume a batter's stance beside them. "Clarify yourself, hero," he'd say, "or you get the bat."

In my mind, Colonel Fontenot stood halfway between my commander at Fort Drum and Petraeus. Rather than wear the traditional Battle Dress Uniform, he donned tanker-crew overalls. You'd think he was an crusty old staff sergeant from the motor pool, until you saw the eagle insignia that denoted him a colonel. He quoted Thucydides and Voltaire the way most people might talk about the weather. He wasn't pretentious about it—he'd be chain-smoking Marlboros the whole time—and he could easily transition to pop-culture talk that the youngest and greenest soldiers could relate to. He could also be wonderfully profane.

Fontenot was not a physical fitness fanatic like Petraeus. Fontenot would often say, "I don't give a rat's ass if my soldiers can run two miles. All I give a fuck about is how fast can they reload tank rounds and how quickly can they change a broken track." Fontenot had led a tank battalion during Desert Storm and, while they don't award CIBs to soldiers in armored units, he saw more combat in those 100 hours of the First Gulf War then most infantrymen did in their careers.

Greg Fontenot was a goddamn killer.

Fontenot said that only one of us—Mark or I—could be in the meeting with him. Mark was the primary planner, so naturally he stayed. I left the tent about the same time the Serbs arrived.

There was a bombed-out school nearby; another officer and I went to look around while the meeting was going on. There wasn't a single piece of glass in the windows. The cold January air filled every inch, turning the colors blue and reminding me, in each turn, of the emptiness after apocalypse.

In one room, the ground was scattered with crayon-colored drawings, the usual kind young kids make in grade school. Except, when I picked one up, I saw that the child had drawn a house completely engulfed in flames. Green blobs with rifle barrels—tanks, I assumed—lined the background. Stick figures formed a line out of the burning building; one of the stick figures was on fire, scratched over with red and orange and yellow. Another stick figure was drawn in the yard beside the house, laid horizontally, with red crayon coming away from its mouth in drops. I looked at the other drawings, all made by kids who couldn't have been eight years old, and there was so much jagged red and black.

I gingerly folded the drawing I held and put it in my pocket. I didn't ever want to forget why we'd come here.

When we finished at the school, we came back and waited outside the conference tent, a fair distance away. A couple hours had passed, and we saw the gathering break up. I watched the Serbs exit the tent without mingling or speaking; they just walked to their trucks and left.

Mark came out and I waved to catch his attention. I could tell he was amused about something. He pulled out a cigarette and lit it, then blew the smoke casually away like someone might after sex.

"How did it go?" I asked him as he took another drag.

"Oh, it was beautiful, Fred." He smiled at me and tapped the ash from his cigarette. He was already taking his time telling this one.

"Before the meeting started, all the Serbs were in a corner talking—but under their breath, smirking like they knew something we didn't." He took another drag.

"When Fontenot walked in, the most senior Serb general pointed at him, said something, and all the people around laughed.

"So Fontenot walked over to the group of Serbs...." He took another drag. "And they are still talking Serbian with their little smiles. Fontenot was just smiling with them, nodding his head, laughing when they laughed. He was just playing their fool, you know. Finally, he said in English, to the senior one, *excuse me, General.*

"The Serbs stopped talking, surprised that he had the audacity to speak. Fontenot said to him, *did you see all those tanks and fighting vehicles out there when you drove up?* The Serb general nodded yes.

"Fontenot said, *if you fuck with me, I am going to shove them up your ass.* I swear to God, Fred, that's what he really said. No shit. It was fucking beautiful."

★ ★ ★ ★ ★

There was not a shot fired (with the intent to kill) during my deployment to Bosnia. We had only a handful of casualties, and not from direct fire—but from a handful of the millions of mines in Bosnia. For all practical purposes, no casualties—and this seems to be one of the reasons the Department of Defense doesn't consider Operation Joint Endeavor a war. When an Army actually succeeds in its mission without losing anyone to hostile fire, apparently it doesn't count as war.

Sun Tzu said, "The supreme art of war is to subdue the enemy without fighting." I'm with the Chinaman. Bosnia was a war, but a near-perfect one. I'm sure most of my dead friends would agree.

What was so remarkable about Bosnia was the restraint of the intervening forces—especially given how horrific the slaughter of innocent people. We came together with NATO and with Russia—our nuclear rival for decades—under the common purpose of stopping genocide. My favorite war, easily.

There were military leaders then who lamented the U.S. military's change of position—from a power of overwhelming destructive force to a servant of nonviolent peace. In the ensuing years, our military would undergo something of an identity crisis, one it arguably hasn't yet overcome. But I think that, somewhere in there, the war hawks misunderstood why our intervention in Bosnia stopped the Serbs killing innocent civilians.

Si vis pacem, para bellum—if you want peace, prepare for war. Greg Fontenot truly wanted peace. But Greg Fontenot was a goddamn killer, and he was there to kill if he had to. When he told the Serbs he'd shove his tanks up their ass, he wasn't saying what he *could* do. He was saying what he *would* do—if they fucked with him. And they didn't.

★ ★ ★ ★ ★

Good fortune prevented my early exit from the Army. Colonel Fontenot and several other senior officers spoke on my behalf. Through letters and phone calls, they convinced the members of the committee to keep me in the Army.

Somehow, I survived the unsurvivable GOMOR.

Then, I was promoted to major and selected again to attend the Command and General Staff College. I got married in the fall of 1996. We had a baby girl. Madelyn. I went back to Fort Drum and the 10th Mountain. I felt like I'd finally come in from the cold. I was given a second chance, and I made a commitment that I wouldn't let the Army down again. I wanted to be sure the people who stood by me were proud that they did.

★ ★ ★ ★ ★

Steve Townsend, our battalion commander, was a new breed of infantry officer taking command in the Army. He grew up in the same Army that I had, but his experience in the Ranger Regiment gave him an edge. It was a subtle degree of difference, what separates an Olympic gold medalist from the silver and bronze. He, like other Rangers, possessed an absolute confidence in his combat skills and abilities, in the way that legendary gold-medal Olympians know they can't lose. Soldiering was in Townsend's guts. He was born in the wrong era; he'd have felt right at home in ancient Sparta.

One morning I came into the office earlier than usual and saw that Townsend was already in the office so I stopped by to talk to him. He was changing out of his hunting clothes; his woodland pattern gear was wet and grimy with mud, grass sticking to the knees and elbows. His bright orange vest (the one hunters are required to wear) was covered in blackish muck. It wasn't unusual for Townsend to go hunting before work, but I never knew him to come directly from the woods to PT. It was early November and already freezing cold in upstate New York.

I said, "Damn, sir, I *hope* you killed something. It's freezing out and it looks like you tackled the deer. What happened?"

By now Townsend was tying his running shoes. He said, "Fred, there was the most awesome buck I had ever seen, but I didn't have a shot. So I climbed down from my stand." He finished his shoes and looked up at me—past me, really. He got glassy-eyed when he told stories.

"The buck was rooting around. I started high crawling. I stopped every few feet. I had to crawl all the way down and then up a streambed. The water wasn't deep but, yeah, it was cold." He stopped and then sat back in his chair and put his hands behind his head, replaying the moment in his mind.

"I saw a tree about 50 meters away that would have been perfect for the shot. I started crawling again—and then I snapped a damn twig and the deer took off. I never got the shot."

I said, "You must be pissed. Do all that work and nothing."

Townsend stopped daydreaming and looked at me. I remembered again that I had never hunted—and that he knew as much. In fact, what little I knew about hunting I had learned from his stories.

He just said to me, "Brother, sometimes just the hunt is as good as the kill."

LTC Townsend could be nothing besides a warrior. He was made for the vocation's every need, in the same way Michael Phelps seems perfectly built to swim. If you're with Townsend, you're untouchable.

Townsend brought his knowledge and experience to the 4th Battalion, 31st Infantry Regiment, nicknamed the Polar Bears, and created a 500-man fighting force in his image. Soldiers of every rank started talking like him, even dressing like him. He taught us a new language for combat which even I, with 15 years in the Army, had never heard. Before Townsend we spoke of violence, of action, of the brute force of overwhelming combat power. But now we were instructed to be in a *careful hurry*. We were told that *slow was smooth and smooth is fast* when closing with the enemy. Mental preparedness to kill was a matter of being *green, amber,* or *red*. We had classes on how to confront and overcome fear and learned that the state of fearlessness was achieved through *sharing hardship*. Townsend merged this almost-spiritual quality with the scientific, tactical aspects of close combat—how we must consider the trajectory, angle of fire, and velocity of bullets when attacking the enemy, for instance.

The way he did it, Townsend passed on his most important calling to his soldiers: to apply the skills in combat, to make it real for themselves as it had always been real for him.

That opportunity came after 9/11. Parts of the 10th Mountain Division deployed not long after the first invasion of Afghanistan, and our battalion was among those units. I was the XO, second in command. Townsend told me to push the units out for deployment by air to Uzbekistan, then to follow on with the last company to the final destination of Bagram Air Base.

Townsend, two companies, and a portion of the staff made it into theater—but our airflow was cut off, due to a cap on the number of troops in the war zone. In other words: the rest of the battalion would not, in fact, deploy to Afghanistan.

I had formally broken the news in a meeting that included the NCOs. I then asked all the officers to come to my office.

The first to arrive was one of the platoon leaders. He was a lieutenant, and he was one of the most remarkable young officers with whom I had ever served. He was a college wrestler, an incredibly sturdy guy, the kind who would take packs off other soldiers' backs as the miles of a march wore on. When I reminded him that it wasn't his job, that part of the point was to let them get used to it, he looked at me and nodded and said, "Yes, sir. I know you're right, but I want to do what LTC Townsend tells us to do and *share their hardship*."

The lieutenant came in and shut the door. I asked him to sit down. He sat on the edge of a chair, rested his elbows on his knees, put his head in his hands, and began to cry uncontrollably.

He had composed himself by the time the other officers began to fill the room. They surrounded me looking for answers. Many of them stared off in the distance as if they were in shock. In others, I could see their eyes were filling with tears. I then drew on the only source of inspiration that would quell the fear that none of our training had prepared us to confront and that was the fear of *not* going to war. So, I spoke to them as Townsend would.

I stood up from my desk and looked at the officers, making eye contact with each of them. Those who were in a daze were brought out of it and their attention was directed to me and I said, "Alright, this sucks. We all hate that we aren't currently in the fight with the rest of the battalion. I know it hurts, but we have to pull it together, especially how we talk about this with our soldiers. We have to keep their minds on getting ready if we are called. We can bitch amongst ourselves what a suck fest it is right now, but when we leave this office, we have to put our game faces on. We don't know how long this war is going to last. We are currently off the deployment schedule, but we may be alerted to go at any time and that's how we have to approach the weeks ahead. We need to step our training, especially live fire exercises. We have nearly all the ranges open on post now and plenty of ammunition. We'll have a meeting tomorrow to go over your training plans."

I paused. This, I knew, is what Townsend would tell us, but I was lost on what to say next. I could not find the words he had taught us to address the most devastating emotion we were feeling and that was to be warriors, at home, when our brothers were at war. The lieutenant who had arrived to my office first was standing in the middle of the other officers. He towered over us in both his physical presence and the power of the words that he said next. The lieutenant said, "Pro Patria"—*For Country*, our unit motto. The other officers repeated the phrase, *Pro Patria*. That is what Townsend would have told us. We do what our nation asks us to do, whatever that may be.

Those of us who were left behind would share this ironic hardship together. However, we did not have to endure it long. Less than a month after they deployed to Afghanistan, Townsend and rest of the battalion returned. There was a continued reduction of forces in the theater of operations and they were being pulled back to the U.S. I received calls from the command sergeant major and operations officer as they prepared for their departure. They wanted me to make sure I had acquired the requisite number of Combat Infantryman Badges and also purchased CIBs with a star for those who had deployed to Desert Storm, Panama, or Somalia. The CIB with a star had not been awarded since the Vietnam War for those that also participated in Korea or World War II. There were a rare few who had two stars on their CIB. I would have merited a CIB with a star because I had served in Desert Storm.

I left the battalion before the ceremony where those that deployed would accept their awards. I had received reassignment orders for my next tour of duty. I could have stayed an extra day to attend it, but I chose to leave. It was the most cowardly act of my Army career.

★ ★ ★ ★ ★

My wife, Laura, had been accepted to a doctoral program at Florida State University. There were no jobs open for me in Tallahassee, so I was assigned as an advisor to an Army National Guard battalion in Panama City, which was about two hours away from Tallahassee. Laura would stay in an apartment near school during the week, while our daughter, Madelyn, and I lived in a house in Panama City. It was not an ideal arrangement, especially for Madelyn, but it was necessary for Laura to fulfill her dream of becoming a psychologist. We would make it work.

Not long after I arrived to Panama City in the fall of 2002 I received a call from David Petraeus, who was now a two-star General in charge of the 101st. He asked if I wanted to be assigned to his Division. "Something big is getting ready to happen in Iraq," he said. Rumblings of another war beyond Afghanistan were stirring as the debates over Iraq's possession of weapons of mass destruction heated and connections of Saddam Hussein and Al Qaeda were being made. There was nothing more that I wanted to do than go to war with General Petraeus. However, there was simply no way I could leave given the circumstances. I believed my marriage would most surely end if I deployed. The strain that resulted from the situation at Fort Drum had already taken a toll on my family. I told him that I couldn't deploy without going into those details. Before he hung up he asked, "Are you running from the sound of the guns, amigo?"

I would have rather him have called my mother a whore.

I watched Operation Iraqi Freedom unfold on television screens in hotel rooms from Fort Stewart, Georgia, to this island of Puerto Rico, where I stayed while training Army National Guard units preparing to go to the war. This time I chose to be left behind. Regardless of my belief that it was the right thing to do not to deploy, Petraeus' words were lodged in me like a dagger, creating a wound that was not immediately deadly, but rather bled my soul out slowly and painfully. I thought about my battalion commander who did not go forward into combat with his soldiers during Desert Storm. I remembered how we talked about him behind his back, saying what a shameful coward he was. My greatest fear was being realized. I had become him.

★ ★ ★ ★ ★

I was promoted to lieutenant colonel and was selected to command the 2nd Battalion, 39th Infantry Regiment, the Fighting Falcons, a basic training battalion at Fort Jackson, South Carolina. I assumed command of the battalion from LTC Bill Wood in June 2004. A couple days before our change of command, Bill and his wife, Nanci, had Laura and I over for dinner. Bill grilled thick rib-eyes while we exchanged stories about the units we had served in and people we both knew. It was early June and the heat and humidity that Fort Jackson were infamously known for had reached record levels. The smoke combined with the intensity of the grill's fire and created an almost suffocating mugginess like we were in a Native American sweat lodge.

Bill pulled two bottles of Budweiser out of a large bucket of ice. He handed me one and put the other to his head to cool himself. We then popped the tops and took long pulls on our beers. Music played from the outdoor speakers of his stereo system. Kenny Chesney, Jimmy Buffett, and Jack Johnson provided a beach atmosphere that must have been soothing to Bill, who was from Florida's Panhandle. His head danced in a nodding rhythm to the relaxed melodies as he turned the steaks. Then he said, "This heat will kill your soldiers if you're not careful. I mean, dude, they will literally walk themselves to death." He told stories about how several trainees had been killed from heat exhaustion during foot marches and runs. "This is the only thing you really have to micromanage. You have to be there with them on the days that it gets really hot and you have to walk up and down the line, making sure they are drinking water and staying hydrated." I thought this was an odd thing for a battalion commander to do. While it's expected for the commander to participate in the training exercises to ensure standards are met, the inspection of individual troops is usually left to subordinate-level leaders. However, as I listened to Bill explain the differences of commanding a training battalion compared to a combat unit, one thing became very clear to me. Bill loved his soldiers. It was not the contrived admiration that many leaders would express with the wave of a hand over their formation, exclaiming *I love soldiers* as if they were a newly purchased set of lawn furniture. Bill loved them each individually in the way that he loved his only child, Rachel.

It was cooling off, but still hot even compared to the state from where my family and I had just moved. After we ate, Bill and I stayed outside on his porch while Nanci and Laura went inside to escape the heat. We switched from beer to bourbon. Bill brought out a brand-new bottle of *Maker's Mark*. He cut off the red wax that secured the cap and poured two fingers worth of the smooth Kentucky whiskey. He took out two *Presidente* cigars from his pocket, cut the tips, and handed one to me. We lit up, blew smoke, and sipped our drinks. It had occurred to me that I had not asked him what he was doing next. "They wanted to send me to D.C. to work in the Pentagon, but there was no way in hell that I could live with myself if I did. I worked a deal to get assigned as deputy commander in the 3rd Infantry Division. We'll be deploying to Iraq in a couple months," Bill said casually.

Bill puffed on his *Presidente* without taking it out of his mouth. I looked at him from where I sat. It had gotten dark now. The glow from the porch light mixed with the haze from the cigars and created a surreal image of Bill against the night like he was a jazz saxophonist in a smoke-filled bar. He then said, "I have been *talking the talk* about selfless service to these soldiers for the last two years. It's time that I *walk the talk*."

Over a year later, Bill was in Iraq and had taken command of a battalion after serving as the brigade's deputy commander. One of his companies came under attack from an IED, and Bill responded by maneuvering to the location where one of his commanders had been killed. As soon as he stepped from his vehicle, a secondary bomb exploded, killing him instantly. I went to his memorial service at Fort Stewart, where he had been stationed prior to deploying. I met Nanci and Rachel there to give my condolences. I sat in the back of the chapel with my command sergeant major. They showed a video collage with pictures of Bill throughout his career and played Bill's favorite songs. I remembered one of them from that night on his porch. As photos of Bill rotated across the screen, Jack Johnson sang the words to one of his hit tunes, *where did all the good people go?*

I commanded 2-39 Infantry for two years. We took civilians and transformed them into soldiers during nine-week cycles that culminated with a graduation ceremony of grand proportion where often nearly a thousand newly minted warriors would march across Fort Jackson's parade field. Parents and family members came from across America and filled the stands to capacity. In my speeches, I would turn and look behind me to the spectators in the bleachers and ask those have served in the armed forces to stand and be recognized. Nearly half the audience would most always take their feet. It occurred to me then how much military service was a family business. After the ceremony I would go out on the field to mingle with the soldiers and their parents. It was one of my greatest sources of happiness to see the pride in their parent's faces and have them say *I sent you a boy and you gave me back a man.*

After one particular graduation, I approached two parents and their son. He stood tall in his dress uniform and black beret. I congratulated the soldier and said to the mom and dad that they must be proud of their boy. The mother thanked me and said, "I'm proud of all my boys. His oldest brother is in Afghanistan. He's been there about four months." She then paused. Her hands then went to her lips and chin as if to prevent the words she was about to say from coming out of her mouth. She faltered, seeming to lose her balance for a moment. Her husband put his hand on her shoulder to steady her. The woman then stood straight and tall like her son and said, "We lost our middle son in Iraq two months ago before this one came here for basic training. I'm sure he's watching down on us though. I am so very proud of him. He died saving one of his buddies. They gave him a Bronze Star for Valor." My stomach dropped to my knees. I could not believe what I just heard and said the only thing that came to my mind and told her that I was so sorry for her loss. This woman stood even more erect and it was like she grew larger before my eyes and assumed the posture of a giant towering over us. She then said in a resolute, almost stern voice, "Don't be sorry, Colonel. The nation will now have all my children. It is the least I can do."

Midway through my second year of command my assignment officer called and told me that I was going to be stationed at Guantanamo Bay to serve in the organization that supervised the prisoners of war there. I told that I wanted to go to the war. He said I couldn't. I told him I was going to the war. He said no. I called General Petraeus, who was now a three-star general, in charge of the Combined Arms Command Fort Leavenworth, where he was, among many things, writing the doctrine for counterinsurgency. I asked him to help me get to the war in Iraq. About the same time, Steve Townsend, who was now a colonel and commanded a Stryker Brigade in Fort Lewis, Washington, that was deploying to Iraq emailed and asked if I wanted to go. His deputy commander had been injured, and he asked if I wanted to replace him. Townsend and Petraeus got me assigned to the 3-2 Stryker Brigade Combat Team (SBCT), the Arrowhead Brigade. It was like Petraeus was giving me the opportunity to find redemption for not deploying when he asked and Townsend was presenting the chance to deploy that I had been denied at Fort Drum. Like two powerful magnets, these two men pulled me out of the center of Hell and provided me the prospect for salvation that I thought would never come.

I left Fort Jackson for Fort Lewis on the 6th of June 2006, the day I changed command. The plan was for Madelyn and Laura to ride across the country with me and then fly back to Florida after I got settled at Fort Lewis. We stopped at my parent's house in Illinois to visit with them before proceeding on. Laura and I sat on my parent's back porch that overlooked a lake our first night there. The sun was setting and its orange glow reflected from the lake's surface that was unruffled except for the ripples caused by fish that bit at their prey at the top of the water. Fireflies danced to the chirping of crickets as darkness fell. We sat there into the night barely speaking a word. Finally, I took her hand gently. She looked at me and I said to her, "I have to go the rest of the way alone." She squeezed my hand tight. Her eyes filled with tears and she said, "I know. You have already left."

I drove the rest of the way across the country by myself. I made it to Washington State in two days. I was in Iraq a month later.

★ ★ ★ ★ ★

I had dual roles in Iraq. I was the deputy commander of the brigade, which meant I was second in command and performed roles to expand the brigade commander's span of control. Townsend would send me to locations where he could not physically be present or to oversee missions that supported tactical operations. I also commanded a provisional organization called the Brigade Troops Battalion, or the BTB. The BTB was unique to the 3-2 Stryker Brigade Combat Team. Colonel Townsend created it to provide oversight of the separate companies that normally do not have a senior leader in charge of their administration and logistics. Those companies included engineers, military intelligence, communications and signal, chemical detection and protection, headquarters and administration, and the anti-tank company. While my position as the leader of the BTB was not officially a battalion command, the fact remained that I was responsible for the health, welfare, and discipline of nearly 550 soldiers. Probably to a fault, I saw my duty as the BTB commander as having a higher precedence of my two roles, even though I spent the majority of my time serving in the capacity of Townsend's DCO.

My position required me to move around the battlefield, so I was provided a security detachment that consisted of three Stryker vehicles that also served as the brigade's mobile alternate tactical command post, or TAC B. Within those vehicles were the men, and sometimes women, with whom I would share my combat experience for the next 15 months.

★ ★ ★ ★ ★

I was standing on the corner in a neighborhood within Doura, a borough in South Baghdad. It was not far from where Bill Wood had been killed two years before. I was with two other soldiers, Staff Sergeant Jared Knapp and First Lieutenant David Abuchallak. Suddenly, we heard a snap, followed by several others. We could see dust kick up in the street. Someone was shooting at us. We looked to try and identify the shooter, but we couldn't locate where the firing was coming from so we did what any good soldier would do: we ran like hell. We sprinted along a wall that lined the street. The bullets impacted behind us, and I could see specks of concrete explode with the impact of each round. They were quickly gaining on us.

We saw a doorway a few yards ahead of us. When we reached it, Knapp pushed me and pulled Abuchallak in just as the bullets caught us and passed by down the street. We tumbled head over feet and landed tangled with one another. As we gained our bearings I felt something running down the back of my neck. Abuchallak came out from under my leg disheveled with dark ooze dripping from his mouth and chin. Knapp had landed a few feet beyond where we were sitting. He was checking himself for wounds. Then we noticed 20 sets of eyes surrounded us. Finally, one of them, after what seemed an eternity, bleated "Baaah, baaah." We had landed in a goat pasture in a pile of goat shit. Knapp, Abuchallak, and I looked at one another and started laughing. We didn't stop for a long while.

Each engagement with the enemy where our team emerged unscathed from the encounter was a celebrated event that was regaled upon when we returned safe home to our forward operating base. After a formal debrief of the day's events and the cleaning of our equipment, we would gather at our subdivision of CHUs, the containerized living quarters where we lived when not out patrolling or on extended missions. Our operating base, FOB Liberty, was divided into sections of large organizations like infantry, artillery, and logistics support battalions and brigades that consisted of thousands of CHUs and created neighborhoods like New York City's Harlem, Chinatown, or Little Italy. Smaller units like ours would stake out their city block and designate stoops, on the steps of CHUs, where troops would gather. We were tribes; I was the chief of the TAC B nation and given the esteemed title of *Uncle Fred* by my warrior clan. That moniker was a private one that was used only amongst the members of TAC B. All other times I was Sir or LTC Johnson. I preferred Uncle Fred though.

Specialist Michael Martin, a Stryker driver and sometimes .50-caliber machine gunner, had procured hamburgers and hot dogs and was firing up his grill as we began to filter over to the TAC B area. Martin was from Florida and was older than the rest of the soldiers. He had assumed the duty of chef and proclaimed himself Master of BBQ, dispensing his wisdom of grilling like Bubba would about shrimp to Forrest Gump. I was one of the last to arrive to Martin's stoop. The smell of charcoal and lighter fluid was thick. Flames rose high from the grill like pointed peaks of a bright red mountain.

The members of TAC B sat in a circle around Martin, but far enough away not to be consumed by the heat of the fire. Some had lawn chairs, while others had created seats from MRE (Meals Ready to Eat) cases or anything else they could find that was somewhat comfortable. When I walked up my driver, Specialist Bryan Henning, was strumming a guitar he borrowed from Brian Kerrigan in TAC A, Colonel Townsend's security detachment. Henning was from Chicago and was wearing a Blackhawks jersey that he often donned at our post-mission galas. He was sitting next to Specialist Rod Rodriguez, who had been my driver but had recently switched to another vehicle. I pulled up a couple boxes of MREs and sat down with them.

Sergeant Joey Cambry, my .50-caliber gunner was sitting next to Specialist JD McClure and Dennis Riehle. They were triple-teaming on the story of the day.

"First the L.T. comes into the truck looking like he just ran a marathon or something," McClure said of Lieutenant Dave Abuchallak, my interpreter. "Then comes Knappy and Uncle Fred, and they both get in their hatch and don't say nothing."

Cambry was drinking a Rip It, the energy drink that we kept by the caseload in our vehicles. He added with a disgusted shake of his head, "The boss gets up in his hatch and I'm sitting down in the gunner's seat and I notice he's got this thick dark mud all over him. His ass is right up next to me, and I smell something. I thought maybe he crapped his pants." Henning stopped playing his guitar and chimes in, "Yeah man, I could smell it all the way up in the driver's hatch. It stunk like hell."

Knapp was standing up next to Martin who was putting hamburgers on the grill. Knapp acted like he wasn't listening, but finally had enough and said, "Fuck you, guys." He then told the story of our foray, embellishing the danger we faced and avoided any discussion of our near capture by the ferocious goatherd. Martin, turned the burgers, stepped back from the smoke and wryly commented, "Yeah, you guys were literally in *the shit*, huh." Knapp flipped Martin off and told him to get back to the woman's work of cooking.

Staff Sergeant James Jastrebski, held an open bun as Martin slid a steaming burger on it. Ski had been the TAC B platoon sergeant. However, he was now with the 18th Engineers, one of the companies in the BTB that we often patrolled with during our circulation of the battlefield. Ski grabbed some ketchup packs from a table we had filled with condiments taken from the mess hall and remarked, "What's the deal with TAC B and bodily functions? Remember in Mosul when that captain pissed himself."

When we first got to Iraq, I had implemented the practice of taking members of the BTB, who were normally confined to the FOB, on patrol so they could experience the war from a personal perspective. I called it a test ride. I did this for a couple reasons. First, I wanted them to understand how the work they performed in the FOB was significant to the overall mission. I found, all too often, that those who did not engage in direct combat operations felt their jobs were not important, which was the furthest thing from the truth. They were chastised for being Fobbits, or what the Vietnam-era veterans called *REMFs* (Rear Echelon Motherfuckers). I despised both terms and forbade them from being spoken in my presence. However, the biggest reason was far less practical and dangerously philosophical. I wanted them to be able to tell a war story when they returned home. I wanted, if they were ever asked, *did you ever see action?* for them to authentically be able to say, *yeah, I got outside the wire.* The captain that Ski was referring to was on a test ride.

The attack happened quickly as was the case with most IED ambushes. The explosion of the bomb was violent and every vehicle in the patrol felt the shock of the blast. A rocket propelled grenade passed by the back of my vehicle and exploded against a nearby wall. We returned fire, and the insurgents disengaged. We then maneuvered out of the kill zone and got into a position to locate the enemy. The members of the dismount team assembled quickly to exit the Stryker. We needed two people to stand overwatch in the vehicles. Sergeant Dey-Dey Wise was one of them, but we needed one more. I looked at the captain, and he was literally shaking. His feet and hands twitched. His weapon was between his legs and then fell to the floor of the vehicle. I pointed and told him to get up into the hatch. He wouldn't move. I then said, *Dude, you have to pull security to cover our movement. We got no time to fuck around.* He got up into the hatch and covered us as we entered a house to see if the enemy was there. The building was clear. When we got back, the captain had urinated all over himself, but he was still pulling security.

Dey-Dey sat back in a lawn chair with a plate full of food on his lap. He wore a sand colored ball cap that featured our brigade's emblem on its front panel. We were in the Arrowhead Brigade, and our patch consisted of an embroidered image of a Native American in a full headdress that was stitched in red, white, and blue and set in the center of a white star that stood out against a black shield-shaped background. Dey-Dey was Seminole Indian and proudly wore the hat without any sense of political incorrectness; he severely chastised anyone who made claims that it was an inappropriate symbol of our unit.

Dey-Dey took a bite of a hot dog and said, "Yeah, I have to give that captain some props though. He was scared shitless, but he stayed in that damn hatch. That was a hell of an introduction to Mosul for a guy who had never been off the FOB."

Sergeant Shawn Sprahler, a .50-caliber machine gunner, sat across from Dey-Dey and said, "It was nothing like the time in East Baghdad when we brought that new lieutenant out, and we got hit by that fucking EFP. You want to talk about *say hello to my little friend.* That was a motherfucker."

Sprahler was referring to the explosive formed projectile IED that had just been introduced to the battlefield. The EFP was particularly lethal. It could easily penetrate the armor of a tank, so it pierced the thin protection of our Strykers like a knife through hot butter. It had become the most feared weapon we faced. The sound of an EFP was like the clap of lightning striking right next to you. It was a reverberation that shook your soul as much as it did the ground around which it struck.

On the day that Sprahler was speaking, we were returning from a meeting with the Iraqi security forces. It was a rare time the entire command group was in the same convoy. Command Sergeant Major Jeffery Du was in the lead vehicle, followed by Townsend, then my Stryker and the brigade operations officer's truck were last in the order of movement. We were approaching a solitary refreshment kiosk in the middle of an isolated city block. Usually local residents would have filled the streets and sidewalks, but none were in sight. Soft drink cans were placed in neat rows on the stand's shelf. However, no one was present at the stand to sell the items. It was an anomaly that signaled trouble but before we could react the operations officer's vehicle was hit by the EFP that was hidden behind the kiosk.

Smoke billowed from the disabled Stryker. The hatch had to be manually let down and when it was finally open, soldiers emerged from inside like ghosts in a mist-filled cemetery. The smoke was from a signaling grenade that had exploded. The faces of the occupants were coated in a white a film that reminded me of pictures of people who were the World Trade Center towers on 9/11 and were covered by the dust from the buildings' collapse. Staff Sergeant Paul Ythemar, a member of TAC B, was one of soldiers.

Ythemar was sitting in our circle next to Sprahler. Martin had finished cooking, so he fixed a plate of food and sat on the other side of Ythemar. Martin had survived a couple IED attacks. He put his hand on Ythemar's shoulder. Ythemar was a huge man, originally from Saipan and a magnificent soldier. He had a ferocious appetite, but he stopped eating and picked at his food when the topic of conversation switched to the experience he was intimately familiar. He stopped eating altogether and just looked at his plate not saying a word. The detonation of the EFP was so close that it passed through the vehicle from one side to the other. The projectile actually penetrated between where Ythemar and Staff Sergeant Alberto Santos, our medic, were seated. Neither one was hit. However, not everyone was as lucky. Shrapnel struck the S3 operations officer, who manned the same position in his Stryker that I did in mine.

Our group sensed the sensitivity of the subject matter to our comrade who barely survived the attack. It was like when a joke unintentionally crosses the line of politeness. Staff Sergeant Matt Hudgeons, TAC B's platoon sergeant, quickly changed the subject to the events immediately following the attack.

"So we drop the back hatch and smoke comes out like something out of a Cheech and Chong movie. We can hear the S3 screaming *I've been hit*. We pull him out and Tienda is checking for wounds," Hudgeons says, referring to Sergeant First Class Carlos Tienda, the medical platoon sergeant. "We are all trying to calm the S3 down and all of the sudden he gets in a real panic and says *check my balls.*

I think my balls were hit." Hudgeons takes a sip of a non-alcoholic beer, steps back, and waves his hand like *no way* and says, "I wasn't going to grab his balls so Tienda reaches inside his pants and squeezes the S3's nutsack and tells him *it's all intact there partner.* After that, he chills like nothing else matters."

Hudgeons' story lightened the mood and changed the conversation away from war stories to news from home, sports and other musings. The fire from Martin's grill smoldered down, as the charcoal turned gray until there was only a hint of red embers. Martin's extinguishment of the fire with a bottle of water signaled the closing of the pow-wow. Henning strummed his guitar and hummed a tune he was making up. Dey-Dey napped in his chair with his cap over his eyes. Rodriguez read a *Men's Health* magazine in the dwindling moments of sunlight that were left. Other members of TAC B departed in ones and twos to their CHUs where they would watch reruns of "The Sopranos" on their computers until they fell asleep. The next day would bring a new mission and another gathering of the TAC B tribe.

The story of Ythemar's near-death encounter and the wounding of the S3 in the EFP attack was the closest time I remember ever talking about events where real tragedy occurred, and even then it had to be masked with humor. The S3 would recover and even return to theater, but the EFP blast would leave invisible wounds and memories that long endured.

We especially never discussed those times when the evil nature of war was revealed to us in images that were too dark even for our combat-toughened compassions to entertain. For example, we never talked about the time when we were on a patrol and saw a boy throwing rocks at something in a field next to his house. Rocks sometimes turned into grenades, so we were cautious as we approached him. The young man's father came out of the house and hurried him inside, but remained on the sidewalk to greet us. We asked him what the boy was throwing rocks at and the man pointed out to a bloated corpse. *Bodies are everywhere*, he said in Arabic. Later, 17 more bodies would be found just down street at a monument named the Children's Memorial.

We did not have discourse about the time we would participate in the liberation of a man from the basement of a house that was used for executions. The man exited his dungeon in complete shock. When we went to examine the room where he had been held at least a dozen blood-splattered marks were lined in a row at varying heights. The executioners had drilled holes in the heads of their victims with a power tool.

Nor did we ever have conversation about the day after we attacked a town and went in to assess the damage and an Iraqi women in a full black burka wailed and asked us to explain why one of our bombs meant for a terrorist cell next door had destroyed her house. The woman then dropped to her knees, clasped hands and pleaded that we help dig her two daughters out from underneath the rubble.

And the day Sergeant Freeman Gardner was killed was absolutely never mentioned during our evening ritual.

We would remember our fallen at formal memorial services where we honored them with rifle, helmet, and boot monuments, 21-gun salutes, and the playing of *Taps*. We would then put their memory away somewhere else in our consciousness. Thirty-six men and women died in the Arrowhead Brigade during our 15-month deployment. There were 35 memorials in theater. I attended about half.

The last of our brothers to pass would be memorialized when we returned from the war. A lieutenant, he had been shot in the spine through the narrow aperture between his MICH helmet and body armor. He had been placed on life support and sent home. He could still speak, but he would never be able to live without the machines that kept him breathing. He was 25 years old and a great athlete. He delivered his own eulogy in our chapel at Fort Lewis through a tape that was made before the plug was pulled at his own request.

When Churchill said, "There is nothing more exhilarating than being shot at without result," he failed to mention the feeling when there is a consequence of being engaged by the enemy and it ends in death, mutilation, or something worse.

That was my war in Iraq. And that's all I have to say about that.

★ ★ ★ ★ ★

Not long after my return from Iraq in 2007 I was selected for promotion to colonel and command of a brigade. I was also chosen to attend the United States Army War College in Carlisle, Pennsylvania. I was united with my family in Carlisle after two years of separation while I stayed at Fort Lewis after the deployment during my wife's completion of her doctorate and internship in Florida.

I assumed command of the Accessions Support Brigade in July 2009. I was put in charge of the most unique command-select brigade in the Army—there was literally only one of them. I was stationed at Fort Knox, Kentucky, but the units I commanded were all over the globe. My brigade was responsible for helping tell the Army story and promote recruiting. It consisted of the Golden Knights, the Army's premier parachutists; the Army Marksmanship Unit, the best shooters in the world; and, the Mission Support Battalion that had a fleet of trailer-tractors with state-of-the-art event-marketing technology designed to inspire young men and women to enlist in the Army. Nearly all the soldiers in my command were senior non-commissioned officers who were specially selected because of their rare abilities to either shoot, skydive, or recruit. While not remotely a combat unit, the brigade had an important mission, one worthy of a capable leader. Unfortunately, they were denied one when I took command.

★ ★ ★ ★ ★

The Fallen Heroes Project commissioned artist Michael Reagan to hand-draw portraits of armed force members who died during their service in Iraq and Afghanistan. The drawings were then given to the families of the fallen at no cost. The sketch of Sergeant Freeman Gardner displays the face of a young African American man in fashionable wire-framed glasses that are barely noticeable except for how they draw your attention to his eyes, ears, and nose. His eyebrows are perfectly arced around the top of the eyeglasses. Those features highlight Freeman's grin. His smile has the appearance that it is a permanent expression, one that he carries regardless of his mood. The impression you immediately get is that he is much wiser than his age. Those who never met Freeman are left with the disappointment that they will never have a conversation with him and learn what was behind the grin. I was one of the fortunate people who had that pleasure to know him.

The Fallen Heroes Project website was saved as a favorite on my computer at my office, where I had started coming early. My sleep had become broken into intervals of hours and then minutes until it became senseless to remain in bed and stare at the ceiling, even though I preferred that option to when my eyes did close and I would drift off into a slumber. At work, I would click onto the site and look at the sketch of Freeman. I would force that image into my mind to replace the one of him lying dead on a gurney at FOB Liberty in Iraq, where the lips that made up his grin had turned purple. But the grin remained, even in death. He looked as if he had been tucked to sleep with a wool blanket that covered him to his chest. The coarse linen concealed the trauma caused by the explosion of a 122-millimeter mortar that propelled shrapnel into his torso. However, the only visible sign of injury was a small cut just above his perfectly arced eyebrow. My obsession with that seemingly trivial abrasion possessed this particular dream.

There were other apparitions that visited me in my sleep. The face of Joe Fenty was one. It was from a professionally done photograph taken when he was a battalion commander and still alive. The red stripes of the American flag behind him blended with red cheeks that appear sunburned. He looked as I remembered him after a trail race in upstate New York after we ran together on Virgil Mountain, only then the brightness in his face was from running for three hours.

There was also the woman on her knees pleading for us to help dig her daughters out from under a building we destroyed. There were the bloodstains on the wall of the death chamber and bloated bodies. There was even the dead Iraqi truck driver from my first deployment to the Gulf with flies and gnats circling around his lifeless eyes.

But what brought me out of my sleep on most nights was the snap of a single bullet. It was not the crash of a barrage of gunshots as we experienced in firefights or even IED explosions. It was the sound of one solitary bullet that I actually heard in my mind that awakened me.

There was at least two times I was engaged with one round from someone I assumed to be a sniper. Both instances I recall were in Mosul. The first round just passed by me leaving no effect other than the noise it produced. The other round seemed to have been fired from right next to me and slammed against of the ammunition container of our .50-caliber machine gun. I am not sure if the intended target was me or Staff Sergeant Hudgeons, who was in the hatch next to me. Nevertheless, it left a dent in the steel box that looked like it had been pounded by a sledgehammer. I'm not certain which of the instances it was that I dreamed about. There was never any imagery associated with the sound. It was just the snap that would bring me out of a dead sleep.

One other vision had become more and more prevalent in my sleep, and that was of me putting a gun to my head. I never pulled the trigger, but the image of me placing a pistol to my temple and holding it there returned night after night.

The effect from the disruptions to my sleep was made worse by an increase in my drinking. My taste for alcohol had evolved from a love of beer to an even greater affection for bourbon. More often than not, I drank both in excess to get me to the point where I would pass out and drift off to a short but relished period of blackness where no dreams would come. However, when Freeman and the others revealed themselves, and the *snap* eventually came, I would be startled awake unrefreshed and dull. I would carry the weariness with me to work where I would click on the picture of Freeman and then try to run the melancholy out of my system on the dark streets of Fort Knox.

In those moments before sunrise, I would run along the shadowy corridors of the Army garrison past the historic homes and parade field to the barracks area where basic training soldiers were starting their exercise regime. Drill sergeants barked out orders and called cadence as the trainees shuffled in columns and repeated their instructor's lyrical lines in a rap-like rhythm. I would follow behind them and sing along in mind:

Up in the morning too soon; Hungry as hell before noon.
Went to the mess sergeant on my knees and said
Mess Sergeant, Mess Sergeant feed me please.
Mess sergeant looked at me with a big old grin and said
If you're going to be a soldier you got to be thin.

The pace was very slow, but some soldiers still struggled to maintain the rate of speed. When one would start to fall behind, a drill sergeant would *inspire* them to consider that it was in his or her best interest to catch up. Sometimes I would get too close to the formation and be confused as a lagging trainee and the drill sergeant would try to inspire me, as well. However, when they realized who I was because of the rank insignia we were required to wear on our reflective vests, they would apologize and explain they thought I was one of their *Joes*, a genderless term of endearment used for soldiers. I wanted badly to discard the vest and sneak in among them and be in their herd, to rub elbows, exchange sweat, and be a part of their struggle. I wanted to cheer on those that fought against their physical limitations just to be a member of the team. Instead, I would break off from their midst and return to my office. I would finish at a sprint to conjure up enough endorphins and adrenalin to get me through the rest of the day.

Of all the ranks I attained in the Army, I liked colonel the least. Maybe it was because I did not believe I deserved the status since I had been given it after my misconduct with the DUI. Others who committed similar offenses have been passed over for promotion or removed from the Army all together. There was also a more political component to the position that gave me discomfort and worked against my inclination for immediate action and to speak frankly. While those traits are admired and even encouraged at lower grades when the expectation is to get things done in the most expeditious manner possible, they are not characteristics that are desired in a senior leader who must think and act strategically.

However, there was something else under the surface of my self-doubt that gnawed at me like a parasite. I could feel its every bite in the pit of my stomach. The pain it created would build until I could bear it no more. Then, in an involuntary retch, I would spew anger. No one was safe from the rage when it came: not my wife, not my daughter, and certainly not my co-workers. In the aftermath of my fury, I would see its effect on the faces of those who bore the brunt of it. An overwhelming sense of guilt would come over me and I would apologize, vowing that it would not happen again. However, that was a promise I could not keep no matter how hard I tried. The endless cycle of wrath and shame repeated itself time and again until those around me became numb to it and numb to me.

I could not wait for brigade command to end. It was such a contrast to previous leadership assignments, where I had never wanted to leave and held on as long as possible. There was no deliberation on what I would do next. I asked to go back to the war. In a way similar to how I felt before deploying to Bosnia, I had to regain my honor. I no longer recognized myself. I somehow felt I would find redemption if I deployed one more time. Once again, I went to General Petraeus for help. He was ending his tour as the commander of the International Security Assistance Force in Afghanistan. I asked if he could help get an assignment there. He came through again and I was given the position of the senior advisor to the Afghan National Army Chief of Staff, General Sher Mohammad Karimi. I changed command of the Accessions Support Brigade on the 21st of July 2011, and I was in Afghanistan two weeks later.

* * * * *

Afghanistan is a germaphobe's idea of hell.

In a typical morning, I would shake the hands of about 50 people before I sat down with General Karimi for our 8 o'clock meeting. That was no exaggeration. I would have placed my hand over my heart—as a part of the traditional greeting—about as many times. On really good days, I pressed cheeks with three or four people; I hoped that number would keep growing because it meant I had built good relationships.

This ritual in the morning—and its repetition in the afternoon when I left for the day—were the absolutely the most endearing parts of my day. Still, there were challenges with this custom. When estimating the time of any meeting, I learned what I called the "handshake/cheek press calculus." Namely: everyone in the room had to be properly greeted before we could get down to business. That could take up to half an hour depending upon the number of people in attendance.

It was interesting to watch the different levels of participation from various coalition members. A few Americans with whom I worked avoided shaking hands all together. They would say *hi*, wave, and move on. If they did shake hands, they'd grab the Purell as soon as they could and use enough to scrub for surgery.

I thanked God I grew up a hand-shaker. My dad had taught me early in life to always shake hands firmly and look the person in the eye. He told me you could immediately judge a person by how they shake hands; a weak-willed, sweaty-palmed handshake was a sure sign of someone who shouldn't be trusted.

There is something to that, I think. The handshake originated as a gesture of peace, to show that hands were free of weapons. From the days of club-wielding cavemen to the present, human beings had evolved to a point where gestures like handshakes are genetic code; they're a signal from one human to another. Those who refused to extend their hand, or who did it poorly, might have had ill intentions. The Afghans—who had been at war for a couple millennia—had mastered the handshake. It's a matter of survival. There were other gestures and customs, unique to a country and people that offer insight to their intention.

In Iraq, I nearly shot a man named Akmed because he was clean-shaven. In Islamic countries like Iraq and Afghanistan, a man's facial hair is part of his character; a clean-shaved face could be a sign of someone prepared for martyrdom. I thought of Akmed on my walk to work one morning as I observed people we passed walking in the opposite direction. From the greatest distance possible, I would always look first for their hands, and then scan up their bodies to their heads. I would get a little concerned whenever I saw a man wearing a turban. Three weeks earlier, a suicide bomber killed a senior Afghan leader with explosives in his headdress. On the street I would greet people with *Sa-laam Alaykum*, which means *peace be with you*, and I would wait for their reply. I made mental notes of people that didn't respond with *Walaykum Assalaam*, offering peace to me in return. Maybe those people were just having bad mornings, but I had to consider that some of them genuinely might *not* have wished me well.

I would relax a bit inside the Ministry of Defense compound and on the short walk to the General Karimi's office building. Once I got near the entranceway, outside the headquarters, I was inside the "bubble." Most of the guards were commandoes, the very best soldiers in the Army, hand-picked to protect the occupants of the Ministry. The commandoes, many of them Karimi's body guards, ensured our safety because Karimi made it very clear in his orders that no harm would come to the Americans who worked in the Ministry. I hoped that it was because Karimi liked and respected me. However, I suspected that it was for more strategic reasons. If an American was killed, then the coalition allies lose faith in the Afghan abilities to provide security. A loss of confidence could result in fewer resources to help Afghanistan. However, I wanted to believe that the commandoes would protect me because they trusted that I had the most sincere intentions of helping Afghanistan win the war. I told myself that as I greeted them daily in their language and in accordance with their customs.

I loved all the guards, but my favorite was Heytabullah. He wasn't a commando, though. He was the head traffic cop who controlled vehicle movement into the driveway and the drop-off point by the lobby. That sacred spot was reserved only for the Minister of Defense and General Karimi. Heytabullah was the master of that domain. His arm-length red brassards and gold wire-rim sunglasses clashed with his lime-green beret and camouflage uniform, but the piece that fit him perfectly was his whistle. It remained in place on his bottom lip even when he talked. He would be casual and relaxed until he saw the Chief or the Minister driving up. He would then become a frenetic ba ll of traffic-directing energy. His arms blurred into crimson propellers, stopping all oncoming traffic where he stood. His whistle burst into short, sharp yelps, then one long screech as he waved the dignitaries into position. That whistle could be heard from a mile away.

As I approached him from down the driveway, Heytabullah stretched his neck and looked down the road. When he saw me, he shouted, "Johnson! *Cheekgo paz, lappo jap*"—Dari slang for "What's happening?" I shook his hand and pressed his cheek lightly against mine, and we exchanged our usual greetings. Heytabullah switched to English, still holding my hand loosely. We talked a bit longer about how our days were going. Then he said, "Let me walk you to the doorway." We held hands like schoolchildren while we walked past the other guards, all the way up the stairs to the entrance. It was my proudest moment in Afghanistan. I wish someone had taken a picture—but no one would have thought to, because in Afghanistan that's just what friends do.

Heytabullah and I said our goodbyes and I then walked the three flights of stairs to General Karimi's office. When I arrived, LTC Khanullah Shuja, General

Karimi's senior aide, was explaining his military genius to everyone within earshot. In excruciating detail, he recounted one of the hundreds of battles he had fought during his six years of company command. I had heard the story about a dozen times, and in each successive telling Shuja would make his exploits more superhuman. But the ending never changed—Shuja ultimately saved the day with a flash of brilliance and his own steadfast courage.

"I wish I could have talked to General Petraeus about these matters so he could have learned something from my expertise on counterinsurgency," he added, "But since General Petraeus isn't here anymore, I will teach you, Colonel Johnson, so that one day you might be like the great Sun Shuja," he proclaimed with a reference to Sun Tzu, the famous ancient Chinese general and philosopher.

I raised an eyebrow and waited for his great wisdom.

He then said in Pashto: *Kharbooza az kharbooza rang megeerad.* I asked what he just said and he explained, "It is an Afghan proverb. It means *the melon takes the color of the melon.* In other words: the more time you spend with me, the more you will be like me. This is very good for you, for the United States Army, and perhaps for the entire world. Thank you very much." He then resumed facing the large map of Afghanistan on the wall and began a new diatribe on an operation he was concocting in Helmund Province.

I had to excuse myself, having learned more than my mind could ever possibly absorb, and said, "*Tashnob marum.*" Shuja said, "Yes, sir, you go to the bathroom. I will save the ending for when you come back." He turned and resumed talking about his idea, pointing with his finger at some small village he was convinced would one day erect a statue of him. He was oblivious to the fact that Captain Zia, General Karimi's son and junior aide, was playing with his iPad, Major Morris, my deputy, was reading *Stars and Stripes,* and LTC Dawari, Karimi's secretary was half-asleep.

I walked down the hall and greeted soldiers and others who worked there by placing my hand over my heart. I subconsciously carried on this gesture with coalition soldiers back at Camp Eggers. It was a hard habit to break. Weeks after returning from Iraq, I absent-mindedly continued the custom even when walking into a bar for a beer. I saw Major General Gull, General Karimi's executive officer.

I bowed slightly with my hands and arms tightly to my side. This was sort of an indoor Afghan salute. It was most often rendered when entering and exiting the room of a senior officer. When departing, the person of the lower of rank bows, reaches behind himself to open the door, and leaves the room without turning his back to the person most senior in rank.

I had explained this courtesy to other coalition members who did not spend a lot of time with the Afghans. The advice was not received very well. Their response was often that I was *too close* to the people I work with at the Ministry of Defense. I was never sure how to take such a comment. I was always under the impression Afghanistan was our ally in the war. At first I was insulted, like that comment was an insinuation about my loyalty. However, I eventually shook the notion off and took it as compliment instead. I had gone to some length they had not.

I returned to the outer lounge about the time people were filing into General Karimi's office for a meeting. Shuja was still standing at the map. He was reaching the crescendo of his war story. I could tell when he was almost done by his exaggerated hand waves and the speed of his speech. Shuja saw me and said, "Colonel Johnson, I'm near the end. You must hear this." I pointed to the Chief's door to tell him that I had to go to the meeting. Shuja shrugged and continued talking. He did not notice that his officemates had snuck out of the room and left him alone.

I attended almost all of General Karimi's meetings. I was there mostly just to listen through my interpreter, Abdullah. Sometimes, General Karimi needed specific information and asked me to find those details. All the seats in the room were filled with one- and two-star Afghan generals. There were four large comfortable chairs and several leather couches. In front of each seating area there were tables with platters of almonds, raisins, pistachios, dried chickpeas, and chocolates for snacking on during the meeting. The "Chai Crew" of Kareem and Negibullah brought in steaming cups of light green tea for everyone.

The meeting began with an introduction by General Karimi. I alternated eating almonds and raisins and sipping tea as Abdullah whispered the English interpretation of the Chief's comments into my ear. General Karimi finished his remarks and then directed questions to the audience. So began the up-and-down dance common to every meeting with the Chief. It was like Whack-a-Mole without the mallet. It's typical decorum in many armies: when asked a question, the officer quickly stands at attention and gives his response to his commander. However, Afghan officers took this to an extreme. If General Karimi even *looked* in their direction, they would take their feet. I had seen brigadier generals nearly knock over tables hurrying to get up from their chairs.

I had become distracted by the contest of which officer could stand the fastest and answer the quickest. It was habit to them. Still, each time a general rushed to his feet to answer, Karimi waved him to sit down. They hesitantly complied, half squatting, back still erect, with their hands on their knees. Suddenly, General Karimi looked at me and asked a question. I was caught off guard and quickly stood, hitting my knee on the table and knocking over all the plates of fruit and nuts. I spilled my chai all over Abdullah and myself. I bent over to collect the chickpeas rolling around on the floor while still looking up to reply to Karimi's question.

He stopped me. "Fred, what are you doing? Sit down, please. You're acting like an Afghan," he said, laughing.

I considered Karimi's comment for a moment and thought, *mission accomplished*. I redeployed about a month later.

* * * * *

My experience in Afghanistan was unlike any of the other wars I had participated. The battle I faced wasn't directly against an enemy. Rather, it was a contest to build the trust and confidence of my Afghan counterparts. I was an advisor after all. No one takes advice from someone they don't believe has true intentions. Sometimes that was the rub: I was not always sure I was carrying messages and advice to General Karimi that were truly in the best interest of Afghanistan. That did not matter, though. First, I did not have all the information to make such an assessment. Second, and more importantly, it was the job I had been given, no different than a rifleman in an infantry squad who is ordered to *take that hill*. There was nothing illegal or immoral in what I was asked to do. As a result, I had to determine the best way to carry out the mission in the best way I knew how.

The problem was that I did not have any military training that had prepared me to carry out the task. I had to draw on other experiences to instruct me on what to do. They came mostly from my childhood. I spent much of my youth in gyms and on playgrounds where I was the only white kid and had to assimilate into a different culture. I worked in my father's bar and learned at an early age that the best way to sell booze was to be a good listener. My Pentecostal mother, who had memorized most of the Bible, imparted the most important advice on me of *loving my neighbor* and *doing unto others as you would have them do unto you.*

While young men and women were fighting and dying in the mountains and villages of Afghanistan, I was given a different hill to charge. Ironically in this war my success or failure, and even my safety, was not dependent on my abilities as a warrior. It was not an M4 or 9mm that protected me and enabled the accomplishment of my mission. It was empathy.

★ ★ ★ ★ ★

I asked to be assigned to an administrative unit on Fort Knox so I could stay in Kentucky and allow my daughter to have four full years in the same high school. I also wanted Laura, who was the director of behavioral health on post, to be able to stay in her position as well. She was doing great and important work for the Army and our soldiers. My daughter had moved eight times in her 13 years. I owed them that.

I was the G3 and director of operations for the unit. It was an assignment I enjoyed because of the people with whom I directly worked, particularly my deputy, Todd Sherman, who was a former command sergeant major. But, like brigade command, I was not particularly cut out for the job. I also thought that my time was simply running out. The sand had been pouring through my hourglass of resolution at a rapid rate the last few years. There were fewer and fewer grains left.

★ ★ ★ ★ ★

The night I send the email to Major General Townsend asking him to get me back into action, I come home reflecting on what I've done. I have a shot of bourbon and chase it with a beer before going upstairs. My wife is already in bed reading. I get in bed and sit up with my back against several layers of pillows that I've propped up. I don't say anything. I just sit there with my arms crossed.

Our dog, a Maltese named Gigi that I got Madelyn after returning from Iraq, crawls up next to me and starts licking my arm. I push her away and she comes back. Then, I just shove her off the bed and say *get the fuck away from me*. Laura shoots a glare at me and finally says, "What in the hell is going on with you?" I tell her about the email to Townsend. She goes back to her book and starts reading again. She turns a page and, as if I have just told her that I have to go take out the garbage, says, "Whatever you need to do."

"Well, thanks for that," I shoot back sarcastically. "What I need is to get back in the fight. Appreciate your support."

At that moment, Madelyn bursts into the room. She has gotten into the habit of waiting outside our door at night. She has long sensed the strain between her mom and me, and they have forged a protective bond with one another to insulate themselves from a toxic demeanor that I have no idea exists in me.

When she storms in, she has the wild-eyed look of a serial killer in a mug shot. Her high cheekbones are flushed red and glow like a volcano ready to blow. She screams at me, "You are such an asshole, Dad. You haven't been the same since you got back from Iraq. I wish it were just mom and me anyway, so go ahead. Go back to war. Leave us alone." She stands with her back straight, but leaning just slightly forward with her fists to her side, as if debating whether or not to swing at me. My young, beautiful, and only child seethes, "I wish I could expunge your DNA from my body." I try to touch her in a soothing way and pull her in for a hug, but she pushes my hand to the side. Then, in complete calm she says, almost in a sigh, "I hate you."

The next day I contact Major General Townsend and say I could not deploy. I tell him I just can't go without explaining why. I do not go to my fifth war. However, what I don't realize at the time is that I'm already fighting it.

NOW OR NEVER

Jail really wasn't what I expected.

Todd, the guy next to me, was in a *Ghostbuster's* costume. He had just performed at a party. After a couple beers he was driving home to Indiana and got pulled over just as he was getting off the Second Street Bridge. Chris was on my other side, and he was wearing a three-piece suit. His tie hung loose around his neck, but other than that he looked like a senior executive. Chris had had some wine after work and was also stopped by the police just as he entered Jeffersonville, Indiana. Tyrone shadowboxed in the corner of the cell. He would do two-minute rounds followed by pushups, deep knee bends, and sit-ups. He adamantly urged us to get up and start moving, "You got to get the alcohol out your system, or you'll be in here all day. Get your sweat on, man." He looked like he knew what he was talking about so Todd, Chris, and I got up and started doing jumping jacks.

Two other gentlemen sat on opposites of the concrete bench we all shared. One dozed on and off. The only time he moved was to vomit in the toilet. He got up, walked slowly over to commode, dropped to one knee, and upchucked for a solid five minutes. Tyrone shuffle over throwing jabs and uppercuts and cheered him on, "That's it. Get it all out. Release the poison. You da man. You da man. Get your puke on." Without breaking stride as he continued to retch, the sickly man raised one arm in the air and pumped his fist, acknowledging Tyrone's praise. The other guy, the oldest of us all, sat with his head in hands, repeating over and over again, "It isn't my fault. It isn't my fault."

Tyrone was the first to get out. When called to the cell door, he bowed to us like he had just won a welterweight bout. Todd was next. He just said, "Take care, you guys." He then handed us his card, which read, "When You Need to Party, Who You Going to Call? Ghost Buster, Inc.

Chris followed Todd. Chris and I had bonded in those hours. He was the only person I told why I had been arrested.

That night I performed a ritual that I had on every anniversary of 9/11 since my Iraq deployment in 2007. I gave a toast to three of my closest fallen comrades who had died in the wars. One shot of bourbon each followed by a beer. I had a Maker's Mark for Bill Wood, because Maker's was Bill's favorite bourbon. Bill died in Doura, Baghdad, on October 27, 2005. Joe Fenty didn't drink bourbon so I drank Woodford, which is my favorite. Joe died on a mountaintop in Kunar Province Afghanistan on May 5, 2006. He was my dearest friend. I had a Basil Hayden for Freeman Gardner because it is so smooth. Freeman died too young on a street in Ameryia Baghdad on March 22, 2007.

I didn't stop with those toasts, however. I sat at the bar alone and continued to drink until last call. I paid my tab and stumbled to my car. I sat on the hood of my truck for a good while. I stared at the keys in my hand like they were a sacred object. I then slid off the vehicle and got in. Putting the keys in the ignition, I said, "Fuck it," and pulled out onto Riverside Drive. I had identified two locations along Riverside, which ran parallel of the Ohio River, where I could drive my car off the embankment into water. I hoped that the impact would knock me unconscious and I would drown. I was driving to confirm which spot was best when I was pulled over by the police.

Chris was as sympathetic as a person could be after learning he was sitting next to someone who was suicidal. But like me he had a wife to whom he had to explain what happened. Worse, he told me that his boss would not be pleased and a DUI could cost him his job. Before leaving I gave him my wife's number and asked him to call her and tell her where I was. He had to memorize it because we didn't have writing utensils. We shook hands, and he left.

When I finally walked out of the Jeffersonville, Indiana, city jail I knew one thing for sure: my wife was going to be pissed.

I called home, and Laura answered on the first ring. I was not allowed to say a word. In an eerily calm voice she said "You're going back to therapy." What wasn't said but was clearly inferred was that, "You go to therapy now, or you will never see me or your daughter again."

Laura had long said that I changed when I came back from Iraq and that I got worse after my deployment to Afghanistan in 2011. She said I had PTSD. Laura is a psychologist, a PhD and the Director of Behavioral Health at Fort Knox, Kentucky, the organization that diagnoses and treats soldiers with posttraumatic stress disorder.

What did she know, right?

It is normal to refuse to sit with your back to the door at a restaurant because you can't see behind you and scan the room for potential threats.

It's normal for the faces of people who died in the war to revolve through your mind like a never-ending slide show of despair.

And, it's perfectly normal to every now and then imagine putting a gun to your head and pulling the trigger.

I did not have PTSD because it didn't exist. You don't get sick serving your nation. I had prepared for war nearly my entire adult life. Moreover, I was an infantry colonel who served 28 years. What kind of an example would I be to junior officers, NCOs, and the enlisted personnel? I would not ever claim that the reason we existed, to fight and win America's wars, could do us emotional and mental harm. I had been to therapy before, but it had nothing to do with PTSD. I had some challenges with anger after I got back from Iraq, but who didn't? Dr. Larry Raskins had gotten me through a bad stretch several year's back. All that was behind me or so I thought.

I went back to therapy because I did not have a choice if I wanted to stay married. And I wanted to stay married very much. And I didn't want to lose my daughter. I didn't return to Dr. Raskins this time though. Instead, I relied on the behavioral health support the Army provided me, which ended up being a full lineup of providers that included two psychiatrists, a psychologist and a social worker.

My sessions with my new team of care providers reminded me how much Dr. Raskins had helped me. The therapy, once again, gave me the cognitive tools to address anger, which was my biggest challenge. However, on one particular occasion they fell short. It was very similar to the last time I did therapy. One day, a fellow colonel made a derogatory comment about a coworker and I lost it. Had not another colleague interceded, I would have hurt the officer badly. I got in more trouble.

My psychiatrist recommended medication. That had been a showstopper for me. I didn't need pills to have self-control. I could do it myself. That's what soldiers do. However, my outburst placed me in another now-or-never moment, one that could decide whether or not I leave the Army with honor.

When I first took the medication the irony was that it was similar to being in a firefight in the sense my world went into slow motion. I could see my anger or anxiety coming far ahead and then my training—the skills I learned in therapy—kicked in. I stopped seeing my friends in that slide show of despair and could remember them at times of my own choosing. Thoughts of putting a gun to my head disappeared from consciousness.

Then one day I woke up after a great night's sleep. My wife and I were in bed side by side, pillows propped, sipping coffee. I reached over and touched her hand and said, "So, this is what it's like to really be normal."

I never plan on returning to jail, but I'm thankful that I ended up there that night instead of at the bottom of the Ohio River. I want to find that cop who stopped me and thank him for pulling me over. I express my gratitude to my wife every chance I get for giving me the ultimatum to seek help. However, those that I am most grateful are my mental health providers who provided me with the training and the tools that gave me clarity to make the right decision in my now-or-never moments.

NEVER THAT COOL AGAIN

I got a call one Saturday morning from a fellow veteran I had met. He said he needed to talk. My friend had served in Afghanistan and took part in some very intense fighting in the mountains of Regional Command—East. Ironically, his closest scrape with death came on his combat outpost when a partnered Afghan soldier opened fire with an automatic weapon on a group of Coalition soldiers. My buddy gave first aid to the wounded in his proximity. Then, he maneuvered to cover, flanked the Afghan, and shot him dead. Point blank. He received an award for valor for his actions. When he got out of the Army, he moved around and landed in Louisville. He was working at a pub that I frequented since I'd retired from the Army a couple months earlier. We had met there along with another veteran who had served in Afghanistan and Iraq, as well. We told stories, laughed, and reminisced with a fervor. One day over a Guinness after a round of war stories, my friend took a sip and said, "We will never be that cool again."

I went to his apartment that Saturday after he called and knocked on the door. He opened it and asked me to come in. He looked very tired. Disheveled. I sat in a chair and he sat on his couch. I had been to his apartment before. I scanned the room, noted a coffee cup full of snubbed cigarettes and a bowl of half-eaten cereal. I saw he had repositioned pictures of himself in Afghanistan and the valorous award citation he had received. They were normally on a stand in an office area of his two-room apartment. Now they were on the coffee table in front of him.

I said, "What's up, brother?" His head dropped. He shook it back and forth a couple times as if he was having an argument with himself. He looked up and said, "The pros outweigh the cons." He then reached under his couch and pulled out a 9mm pistol. He held the weapon comfortably as if it were a part of him. He then lamented, "I just don't want to deal with this shit anymore. Tell me why I shouldn't blow my brains out."

I was not anticipating that question. However, I knew that I could not get the answer wrong. I bought myself time by saying, "You'll bullshitting me, right?" He stared at me, didn't say a word, and waited for an answer. Strike one. I talked about his family, friends, and all the people that loved him. How it wouldn't be fair to them. I felt him soften, but then he said, "They are better off without me. I'm just a fucking burden." Strike two.

Then, I looked at the pictures on the coffee table. He was wearing multi-cam fatigues, full body armor, magazine holders filled, and MICH helmet, and he was cradling his M4 rifle. You could see the mountains in the background, the Hindu Kush, the foothills of the Himalayas. Having been there myself and seeing those photos, I could recall the sparse air at 8,000 feet, taking it full up into my lungs. I could remember looking out over vast valleys and wondering if Alexander the Great had seen the same sights during his failed conquest of this harsh region.

I took one of his pictures and handed it to him. I said, "Put the 9 mil away, brother. We can be that cool again."

He unloaded and cleared the weapon. He locked it in his arms safe and gave me the keys. We got in my car and drove without a destination and did what soldiers do—we made a plan.

MY FIFTH WAR

Wars are not fought alone, even those (especially those) that warriors wage at home within their own soul. When I was at my lowest, that point when I intended to drive my car into the Ohio River, there came together a confluence of guardian spirits, both in people and events, that connected like tributaries into a great waterway and eventually landed me safely on the shore of sanity.

The disgrace I felt after my arrest weighed on me as if I were Sisyphus, a king in Greek mythology cursed to roll a huge boulder up a mountain, only to watch it roll back down, repeating the action for an eternity. The difference was that I never got the reprieve of Sisyphus' walk back down the mountain to retrieve the boulder and begin the pushing again.

While I was allowed to retire and maintain my rank—both of which were in serious question for a miserably long time—I was ashamed of myself. I let my soldiers, leaders, and the Army down. The loss of honor, whether on the battlefield or on a street in Jeffersonville, Indiana, is a deathblow that severs the soldier's spirit. It is a fate worse than death and, more often than not, the modern-day *samurai* opts for *hara-kiri,* rather than live with the shame. Therapy and medication helped preempt such an outcome for me, but the weight of failure was sometimes almost too onerous a burden to bear. Moreover, what weighed almost as heavily on my soul was that I did not understand what was wrong with me.

Two psychiatrists, a psychologist, and a clinical social worker diagnosed me with PTSD. My wife, a psychologist, was adamant that I seek help for PTSD not long after I returned from Iraq in 2007. However, during my last months in the

Army, when the disposition of my discharge and characterization of my service was in question, one Army psychologist claimed that I suffered from a *pre-existing* personality disorder or bipolar disorder. If that were the case, I would either be administratively or medically discharged from the service without full benefits and possibly without my rank and retirement. The rationale for such a decision rests on the belief that the Army is not responsible for the problems a service member had before entering service; therefore, the institution has no obligation to provide certain benefits afforded soldiers who incurred injuries, physical or mental, while serving. The problem with this thinking, of course, is it doesn't explain how the person was allowed into the military if they had a condition that existed before entering the service. There is a screening process that occurs prior to enlistment. It is possible that people lie about a mental health issue. As a former recruiter, I know that happens, but the adverse effects of mental illness should manifest itself early on either at basic training or at some time after they arrive to their first unit of assignment. There is an administrative process for those situations that allows individuals that fit into that category to leave the military without any adverse action. However, my struggle is with service members who have successfully served 10, 20, and even 30 years who are considered for separation for a *pre-existing* condition. That is a very long time to hide a mental illness.

Others within my own circle of senior leaders argued that I should be separated for bad conduct, resulting from my arrest and disruptive behavior, particularly my angry outburst towards another colonel whom I threatened. I recognize there were better ways to handle the situation and I was out of line. However, there remained the question of: *What was wrong with me?*

In 2013, the American Psychiatric Association revised the PTSD diagnostic criteria in the fifth edition of its Diagnostic and Statistical Manual of Mental Disorders (DSM-5). This is the manual behavioral health providers use to diagnose mental health disorders. This description of the DSM-5 criteria for PTSD comes from the Veterans Administration website:

DSM-5 pays more attention to the behavioral symptoms that accompany PTSD and proposes four distinct diagnostic clusters instead of three. They are described as re-experiencing, avoidance, negative cognitions and mood, and arousal.

Re-experiencing covers spontaneous memories of the traumatic event, recurrent dreams related to it, flashbacks or other intense or prolonged psychological distress. Avoidance refers to distressing memories, thoughts, feelings or external reminders of the event.

Negative cognitions and mood represents myriad feelings, from a persistent and distorted sense of blame of self or others, to estrangement from others or markedly diminished interest in activities, to an inability to remember key aspects of the event.

Finally, arousal is marked by aggressive, reckless, or self-destructive behavior, sleep disturbances, hyper-vigilance, or related problems. The current manual emphasizes the "flight" aspect associated with PTSD; the criteria of DSM-5 also account for the "fight" reaction often seen.

I resisted the notion that I had PTSD until one of my therapists showed me this definition and we identified that most, if not all, of the criteria were present in my behavior. With my recognition that PTSD was the problem, I agreed to undergo Prolonged Exposure therapy, a dreadfully difficult experience of reliving specific events that troubled me. We spent the majority of time recalling my role in and reaction to the death of Sergeant Freeman Gardner and the shame that it caused in me. Administered by Cyndie Ramminger, a brilliant and compassionate social worker at the VA, my experience with Prolonged Exposure was a turning point in my therapy. At some point during the eight-week program I found myself letting go of the shame. It was replaced with acceptance that sometimes bad things happen to good people during war.

There remained, however, a struggle that I continued to wrestle down. I could not blame my sickness on the Army and particularly that it resulted from my service to the nation. I rejected the notion with every aspect of my being. Nevertheless, the reality that I came to accept is that blame is an empty emotion. Blame had no place in the personal reconciliation of my condition. Who or what was at fault didn't matter. What mattered was getting better. Part of the process, separate from my formal therapy, was coming to grips with certain aspects of the profession that I chose and the effect it had on me personally.

I volunteered to be a soldier. When I was sworn in as a second lieutenant of infantry in 1985, I thought I knew the inherent risks of my profession. I progressed through years of training that included infantry basic and advanced courses, Airborne, Air Assault, and Ranger Schools, and countless exercises that included the use of live ammunition and explosives. One of my best commanders, Steve Townsend, had professional development classes that specifically addressed how to confront fear and potential death. Another class was on the psychological aspect of killing. Years of foot-marching long distances, training in all types of weather conditions, and learning to live in harsh and unforgiving environments hardened me. When I went to war I was prepared to kill and I was prepared to die. However, during combat something else was asked of me that I was not prepared to face.

Combat changes a person. There is baggage that comes from training to kill and then being asked to use that skill against another human being. The constant state of hyper-vigilance, the recurring exhilaration from adrenalin rush, boosts in testosterone levels, and just the general belief in one's hero status are mind altering, and that state is carried from the battlefield back home. A combat leader also bears the weight of their decisions that result in the death of soldiers like Sergeant Freeman Gardner. Those decisions also cause destruction to non-combatants who live amongst the enemy we want to destroy. The old woman on a street in Baqubah, Iraq, who asked us help dig her two daughters out of the rubble after we bombed her home is just one example. There are many others. The Army prepared me to kill the enemy, not civilians. While it is a brutal fact of war that non-combatants die, that does not make it easy to accept and purge it from your memory once you witness it and play a part.

There is no blame. However, there must be recognition that all the training in the world cannot, with absolute certainty, prevent the mental and moral damage that comes with combat. The Army's Master Resiliency Course, training designed to increase the mental and emotional toughness, could not have prepared me to see Sergeant Freeman Gardner's lifeless body on a gurney or to write a letter home to his mom and live comfortably with the knowledge that I had a hand in his death.

There is no blame. Most importantly, I am not a victim. I was just not ready for it. I waited too late before addressing the effect it all had on me. It took a night in jail and my wife's ultimatum for me to seek help. I found the help I needed in remarkable group of mental health professionals. Dr. Sue Bentley, Dr. Scott Eader, Dr. Karen Grantz, Dr. Ellen Knox, Dr. Larry Raskins, Mark Thurmond, MSSW, and Cyndie Ramminger, MSSW, saved my life. If one thing is taken from this book by my fellow veterans who are having challenges it is to get help from the mental health services the military provides. Then, repay the services of those professionals by staying healthy.

I would remain in therapy, and I take my medication daily. Both will be a part of my life until it is absolutely certain they are no longer required. I seriously doubt that day will ever come. I have accepted it as a fact of life, just as someone with diabetes acknowledges they must always manage their glucose levels. The diabetic may require medication, and even insulin injections, but there are other things (like losing weight, nutrition and exercise) that contribute to their health, and when applied together, reduce the adverse impact of the disease. I would learn those other life saving measures beyond the medication and therapy slowly as I made my transition out of the military and into the civilian world. However, almost until the very end of my Army career, the characterization of my service as honorable was in question.

My therapist's efforts had little to do with the disposition of my final discharge from the Army. That was handled behind the scenes by Brigadier General Jim Iacocca, an officer very senior in the organization to which I was assigned. He kept those that wanted me to leave the Army negatively at bay.

It is an unfortunate reality, though, that there are other leaders at his rank and much higher who deny the existence of PTSD or that bad things happen to good soldiers when they get back from war. Such leaders assert that no one could possibly have PTSD because *they* don't have it. Those who are diagnosed with PTSD are simply weak or faking it, they believe. These types of leaders are the most lethal enemy we face, particularly in a garrison environment. When these leaders administer justice, they lack the empathy to weigh mitigating circumstances. As a result, they waste the lives of soldiers with a mark of a pen and the comment, *There will be no shit bags in* my *Army*. These leaders separate soldiers from service, deny them medical boards, and administer their own version of military justice until the very soul of a soldier is stricken lifeless.

I met one such soldier at the Veteran Administration (VA) hospital. I'll call him Ralph. Ralph was the son of a coal miner in the Appalachia region of Eastern Kentucky. Ralph told me, "I was not going to die under the ground like every male in my family had for the last 100 years. Nope. I joined the Army as soon as I was old enough. I was going to be a soldier. I was going to make a career out of it." Ralph deployed to Iraq. His wife, also a soldier, did not deploy. Ralph was severely wounded and was evacuated to Landstuhl, Germany, for treatment and to recover from his injuries. He was there nine months.

During that time, a fellow soldier that Ralph considered a friend raped his wife. Ralph didn't find out until he got home. Ralph said to me, "When my wife told me what happened, I started making plans to kill that rapist motherfucker. It's all I thought about." Ironically, like in my situation, Ralph's plot to murder the soldier was interrupted by a DUI arrest. Ralph was a sergeant E-5. He did not have my years of service and rank. He didn't have a leader, like I had, to watch out for him. Ralph was denied a medical board, even though both his physical and internal wounds had left him disabled. He was separated from the Army under other than honorable conditions. He is still fighting for his VA benefits. Ralph couldn't find a job because of the characterization of his service. Ralph is now in school studying to be a social worker with a specialty in military affairs. He says, "I ain't going let the son-of-a-bitches do to other soldiers what they done to me."

I was a colonel. Ralph was a sergeant. How do these leaders treat a private? That is a question our military must answer.

I honestly do not know what Brigadier General Iacocca did to help me. He will not discuss the details. All I do know is that I was given a second chance in the waning months of my career. Those last months, even though relieved that I would leave the Army with my benefits, weighed heavy—mainly because of self-pity—because I had ruined a reputation I had worked hard to build for almost three decades. I believed that I was the *shitbag* many people called me in whispers in the offices and halls of the building where I continued to work.

Yet again, I was lucky to have someone willing to help share my load simply by being a good friend. Todd Sherman was a co-worker and a retired command sergeant major who maintained the non-commissioned officer ethos even as a civilian. Todd Sherman is as big as a bear. I had come to call him *T*, not so much because his name started with the letter T, but because that's what the underbosses called Tony Soprano in *The Sopranos*. As far as I was concerned, and pretty much everyone who knew him, Todd could have easily played the part in the miniseries without having to act. Todd took care of me like every good CSM takes care of their boss and that was by providing sage advice. One day Todd came up to me and put his arm around my shoulder. He pulled me in tight like Tony Soprano would his consigliere, Silvio Dante. He put his mouth next to my ear and whispered, "Get your head out of your ass." You did what Todd told you do without question. I would not be sleeping with the fishes so I pulled my dome from the darkness of my buttocks and I made a decision that would become my mantra: *My second act will be better than the first.*

I would be the playwright, producing and artistic director, and the main character in a grand performance of personal re-creation. I knew one thing for sure when I retired from the Army in the fall of 2014. I would continue to serve my nation in some capacity. Service feeds the soldier's soul. There is no diet on earth more satisfying and wholesome for a warrior's spirit, especially one that has departed the military, than being a part of something larger than oneself again. My first helping of that soulful supplement was served up to me with glorious hours spent in one of Louisville's worst neighborhoods.

Even before I retired, I began a search for volunteer opportunities and I found YouthBuild Louisville. YouthBuild is an oasis that rests in the heart of Smoketown, a square-mile neighborhood that is representative of every problem faced by the underserved and neglected populations of large cities throughout our nation. Lynn Rippy, YouthBuild's executive director, and her team created a sanctuary where Louisville's at-risk youth could begin a journey to realize the potential that was denied them in the crime-ridden confines of their respective city blocks. They could learn a trade, earn their GED, be paid a stipend, and be positioned for jobs after their successful completion of the yearlong course. There were only four conditions: there would be no unexcused absences, they had to complete their classwork, they were required to perform volunteer hours, and they would remain drug-free and be given periodic urinalysis.

I contacted Lynn and requested an interview for a volunteer position. Lynn asked what I had in mind. I told her I'd like to be a mentor to the kids, specifically providing military physical training like the "boot camps" that were currently the trend. I would come every morning and put them through exercises. In that process I hoped they would learn to value fitness, gain some mental toughness, and bond together as a team in shared hardship. Lynn liked the idea and asked only one thing before I would start the next day. "Love them, Fred," she said. That was the easiest thing ever asked of me.

At 8 o'clock the next morning I was introduced to the group of 20 young men and women. The social service coordinator, presented me as Colonel Fred Johnson. One of the girls raised her hand and asked, "You mean like Colonel Sanders? He don't look like no Colonel Sanders." I could barely stop myself from bursting out in laughter, but from that point on I was referred to as the Colonel. It was the first time in a long while that I felt proud of the title. Every time one of the kids would say, "Good morning, Colonel" a feeling would come over me like when the sun comes out on a bitterly cold, cloudy day and the beams of sunlight wrap their warmth around you like a wool blanket.

After the introductory meeting I instructed the kids to be out on the school grounds in five minutes. As they sheepishly meandered to form a semicircle around me, their arms crossed and attention diverted everywhere but to me, I made a decision that I would change my original plan of easing into the program to one of shock and awe. I was never a drill sergeant, but they had trained me and I went into *Full Metal Jacket* mode, less the profanity and insults.

For the next thirty minutes, we practiced falling into formation, did pushups and jumping jacks, and ran in place. Near the end of the session I taught them an old military running cadence about jumping out of a C130 aircraft during airborne operations.

C130 rolling down the strip,
Twenty from YouthBuild on a one-way trip.
Stand up, hook up, shuffle to the door,
Jump right out and you count to four.
If your main don't open wide,
You got a reserve by your side.
If that one should fail you too,
Watch out below I'm coming through.

We practiced it a few times and then took off jogging around the block singing it along the way. Neighborhood residents came out of their homes to watch and wave. A homeless man joined us for a few hundred feet, and then abruptly stopped out of breath and shouted to us, "Go ahead with your bad selves!" We repeated the cadence again and again. As they learned the words and felt the rhythm of the rhyme, like a rap song, they began to clap in unison and sing louder and louder.

We marched our way back in through the gates of the YouthBuild compound and formed up for closing remarks. Several of the kids were noticeably exhausted; others dropped and did pushups without provocation. I handed out water bottles from the back of my car, explaining the importance of hydration. I told them to fall in and they assembled roughly in their assigned positions. *Not a bad first day*, I thought. I then asked them if there were any questions and nearly every hand went up. I chose the girl who thought I was Colonel Sanders, and she inquired with all seriousness, "Colonel, what's a C-130?" Others followed in rapid succession: "What's a reserve?" "Why we got to count to four?" "What do you mean stand up, hook up shuffle to the door?"

I looked at the kids and said to myself, *Fred this is not the Army. I may need to try something else.* Then, I reconsidered and thought: *It is better than the Army. This is the best job I've ever had.* The problem, however, was that I wasn't getting paid. I had to find employment to supplement my retirement pay, not to mention my volunteer work with YouthBuild didn't fill up my day. My wife, Laura, wisely urged me to find a full-time job. She told me, "You can't have too much time on your hands. Devil's workshop and all that." I did not have to do a search. An opportunity came to me from one of my oldest and dearest brothers in arms, Anthony Coggiola.

Cog and I were college classmates. We had dorms adjacent to one another. The first time we met he wanted to beat my ass for playing my stereo too loud. The second time we met he came to me with a hand he sliced in a fistfight. He knocked on my door. I opened it cautiously not knowing his intentions. He had a bottle of Jack Daniel's, a spool of thread, and a needle. He walked in and said calmly, "Dude, sew me up." It was cut deep, nearly to the bone. Blood was dripping everywhere. I said, "You're bullshitting, right?" "No," he said, "Sew me up. If the football coach finds out, I'll get kicked off the team." I sewed him up. We became best friends. We graduated from college, and the next time I saw him Cog was in the Saudi Arabian desert as we prepared for Desert Storm and the invasion of Iraq. I watched a cloud of sand billowing in the distance as a convoy of vehicles approached my location. It stopped abruptly in a screeching halt like something out of a Roadrunner cartoon. Out jumps Cog, arms wide open, hands pointing to the sky. He exclaimed, "Dude, Let's go war!"

In similar fashion Cog called me one day and essentially said, *Dude do you want a job?* It was a short-lived endeavor serving as a consultant for his business. However, it bought me the time to actually conduct a job search. Cog will never admit it (because he's that kind of guy) but I believe he sensed that I needed a jumpstart in my transition from the military. He knew about my challenges. Cog responded as a warrior would. He did not leave a fallen comrade.

While working for Cog, I met Tim Peters. Tim was a Vietnam veteran and owner of a very successful construction company in Louisville. Tim became my mentor and, most importantly, my friend. Tim was also married to Lois Mateus. Before her retirement, Lois was the most senior female executive at Brown-Forman and creator of the Woodford Reserve bourbon brand. Lois also took an active role in guiding me through the nuance of relationship building and networking in Louisville's very insular and complex social environment. With very few exceptions, I was blessed with an array of remarkable leaders that facilitated my success throughout my military career. Tim and Lois provided that same level of leadership in my transition to the civilian world. Lois and Tim have been the General Dave Petraeus and Lieutenant General Steve Townsend of my post-Army life.

After several months with Cog, I got settled in my transition and started looking for more permanent work. I solicited the help of Where Opportunity Knox, an organization that finds jobs for veterans, I got an interview and with a very strong recommendation from Lois Mateus I was hired for a position as a development officer at the Fund for the Arts in 2015. The Fund for the Arts is a nonprofit that raises money for arts organizations in Louisville. My work with the Fund for the Arts created even deeper roots in my commitment to the community. However, there was an added value in discovering the vast richness of Louisville's art and culture scene and the benefits it brought to education, wellness, the quality of place, and the economy of the city. It was revealed to me slowly like an orchestra's crescendo or the buildup to a play's climax. With my every interaction with the performing and visual arts, I realized that I, too, was an artist. Not the conventional kind like a painter, musician, or dancer. My art was more ancient. I was a storyteller.

Storytelling was an essential task in my work at the Fund for the Arts. As a fundraiser, it was my responsibility to explain the arts' impact in the city and why a person's donations made a difference. It was the donor's right to know where their money goes. I found the most effective way to accomplish that was to tell stories about people, particularly children, the elderly, and the disabled, who were helped by contributions to the Fund for the Arts. To do that, I had to meet the people, see the positive effect on them, and build the story. I thought I would be a natural at it after years of public speaking in the Army. However, I soon learned there is far more to telling a good story then simply giving a motivational speech. It required the rhythm of a dancer, the painter's eye for the abstract, the musician's ear for subtlety, and the actor's appreciation for the audience. I needed practice beyond the conference rooms and break areas where I gave my presentations. I took improvisation classes and watched Ted Talks incessantly. I started telling live stories at Louisville's *Moth Radio Hour* affiliate, Headliners' Moth StorySlam. It was in Headliners' dark and beer-drenched halls that I began to find my voice as a storyteller. I also experienced the immediate sense of humility when a story is told poorly and the exhilaration of when one is told well.

Many of the stories in this book were told on stage at Headliners. Some did well and got high ratings from the panel of judges. Others were received with polite scores barely above average. However, one story, "Now or Never," placed first at one event, so I was selected to participate in the GrandSlam in Whitney Hall at the Kentucky Center of Performing Arts. That night I told the story live to a packed auditorium. I did not win, but it was well received. It was published in written form, by *Reader's Digest* and broadcast nationally on the *Moth Radio Hour*.

Another benefit of working at the Fund for the Arts was that I got to meet many of the artists, dancers, musicians, and actors. I also became acquainted with the executive directors and staff of the major organizations we supported like the Louisville Ballet, Actors Theater, Louisville Visual Art, Louisville Youth Choir, Walden Theater/Blue Apple Players, Kentucky Opera, StageOne, and the Louisville Orchestra. It was because of the Fund for the Arts that I happened to be present at the Kentucky Museum of Art and Craft when Lee Mingwei, an international renowned artist, introduced the Mending Project, an arts experience that had a profound effect on me.

The Mending Project was an interactive exhibit where sewing and mending serve as a vehicle for sharing. The mender, a volunteer, sits at a table. Visitors bring in articles to be mended, and the pieces remain attached to the threads and piled onto a table to be retrieved at the end of the exhibition. The collective mending celebrates the idea of repair through community and bonding between people. The night before the opening of the exhibit, Lee spoke softly and eloquently about the process when a thought came to me. It was of Sergeant Freeman Gardner. Therapy had helped me come to terms with the guilt I felt in his loss. However, I always felt there was something more I needed to do. After Lee finished speaking and asked for questions, I raised my hand. For the first time ever in an open forum, I spoke of Freeman and the impact of his death on me. I asked to be the first participant in the Mending Project. Lee agreed.

The next morning I brought in a pair of Army camo trousers that I had worn in Iraq. I couldn't be certain if they were the pants I was wearing when Freeman was killed, but I only had a few pair during that 15-month deployment. Nevertheless, these particular pants had tears that could be sewn. Lee gently took them and moved his hand over the fabric and told me he would fix them for me. Lee stitched up the tears with threads of pink, yellow, purple, and shades of orange. I then took the pants and went to a space in the back. I sat alone with the trousers for a good part of the morning. I thought of Freeman, Joe Fenty, Bill Wood, and other friends that I had lost in the war. I considered the seemingly incongruent colors that pieced the tears back together. In that process something else, much more than the pants, was mended.

Thomas Carlyle wrote, "Everywhere the human soul stands between a hemisphere of light and another of darkness; on the confines of the two everlasting hostile empires, necessity and free will." In my fifth and final war, light has prevailed. While necessity and free will have played a part, my battle has been won alongside a legion of kindred souls that encircled me in a time of torment and provided me protection against the fierce foe of darkness.

EPILOGUE

"Once more unto the breach, dear friends, once more."

— **Henry V**, Act III, Scene I —

When I realized that art truly heals, I wanted more of it, and I wanted to bring other veterans along with me. Not long after my experience with the Mending Project, I watched William Shakespeare's *Macbeth* and that notion of art's healing power would be confirmed in the most amazing way.

Kentucky Shakespeare is another organization that the Fund for the Arts supported. The theater's producing artistic director, Matt Wallace, is fond of saying Kentucky Shakespeare is a really social service program that just happens to be a professional theater company. In one of our first meetings, Matt and I began a conversation that would result in a series of life-changing events I never could have imagined.

Kentucky Shakespeare produces the longest running free Shakespeare festival in the nation. Each summer from the beginning of June to the middle of August nearly 30,000 spectators fill the lawn and benches of Louisville's Central Park to watch an array of Shakespeare's plays. However, Kentucky Shakespeare's work doesn't end at the conclusion of the season. In fact, it is just beginning. Members of the company perform in schools and community centers within every ZIP code in Louisville. They also have a program where they perform a two-person abbreviated version of a Shakespeare play in elderly care facilities and local schools. Their work with at-risk kids in Louisville was recognized as one of the most effective programs to help troubled youth stay out of trouble with the law.

Matt Wallace is also the director for a program called Shakespeare Behind Bars, where inmates at the Luther Luckett Correctional Complex perform a Shakespearean play each year. The national recidivism average is 67 percent; the Kentucky recidivism rate is 29.5 percent; and those inmates who have participated in the Shakespeare Behind Bars program have a current recidivism rate of 6.1 percent. In one of our conversations, I asked Matt, "I wonder if a program like Shakespeare Behind Bars could be developed for veterans who have PTSD or those who are struggling with transition challenges coming out of the military?" He answered without hesitation, "Absolutely."

Matt and I then laid out the objectives of the program we would call Shakespeare with Veterans. First, we wanted to establish a safe environment for our members to express themselves with other veterans and address the challenges they were experiencing. Second, we wanted to recreate the conditions that veterans loved most about their military experience, specifically comradery and a higher sense of purpose. Finally, we wanted to connect America's military with the American people by performing in front of live audiences. We wanted to tell our story. The study of William Shakespeare's words, practice of ensemble techniques, and acting exercises would allow us to achieve all three goals. Kentucky Shakespeare's Amy Attaway, the associate artistic director, and Kyle Ware, the education director would lead the veterans on their journey of discovery and healing. However, we had to get funding for the program and we had to find veterans who were interested in participating. Both tasks were easily accomplished.

We raised nearly $15,000 dollars in less than two weeks through online crowd sourcing using a Fund for the Arts program called Power2Give. Over 70 people donated and the Louisville Metro Government matched each dollar contributed with 50 cents. Molly Malones, a local pub where I did a good bit of writing (and drinking of Guinness), along with my mentor Tim Peters donated a considerable sum. The last person to contribute was General (Ret.) David H. Petraeus. General Petraeus took care of me throughout my Army career and he continued to do so when I became a civilian. Finding veteran interest happened at about the same pace with the help of two tremendous veterans.

Al Snyder, a United States Army Reserve lieutenant colonel with two combat tours in Iraq, was also the executive director of the Vet Center. The Vet Center provides behavioral health counseling and VA benefit assistance services to combat veterans. Counselors at the Vet Center are required to have wartime experience as a prerequisite for employment. Lindsay Gargotta, an Air Force veteran, founded the non-profit, Athena Sisters, to provide military women a venue to meet in a community of courage and build empowerment through artistic expression. Both the Vet Center and Athena Sisters had a large population of veterans, and Al and Lindsey allowed Matt Wallace and Amy Attaway and myself to speak to their members about the program.

It was very important to us that the Shakespeare with Veterans membership reflect how we fight our wars. Our intention was to offer the program to the entire fabric of our military, including veterans of every race, gender, sexual orientation, and ethnicity and from as many conflicts as possible. Al and Lindsay helped us achieve that goal by bringing in the first members of Shakespeare with Veterans: seven men and five women from the Army, Air Force, and Navy who had served in the Vietnam War, the Cold War era, Desert Storm, Operation Iraqi Freedom, and Operation Enduring Freedom. Their former military ranks ranged from specialist to brigadier general.

We started Shakespeare with Veterans in February 2016. We met every Thursday from 6 to 8 p.m. at the Vet Center. The sessions started with an opening question asked by Amy Attaway, who became our primary facilitator, with Kyle Ware filling in to help when Amy had other theater commitments. The opening question usually pertained to the military, which helped Amy (who had no previous interaction with service members or veterans) learn more about the armed forces. It also allowed all our members to get to know one another. However, over time the opening question became a kind of therapy session where we could talk to other members of the group about our experiences and things that were on our minds.

The trust gained with the opening question set the stage for the rest of the program. Next, we performed a series of exercises that taught us the basic components of acting and ensemble techniques. These "silly games," as Amy called them, furthered team building within our group. However, they also had a more subtle effect of teaching us how to project our voices, listen more intently, anticipate each other's actions, and improvise—all critical acting skills. Amy would then lead us through readings of selected texts from Shakespeare. We would analyze the monologues line by line trying to determine the meaning of Shakespeare's very difficult language. We learned that Shakespeare was meant to be said aloud and heard, not necessarily read. Listening to his words deliberately in the group setting and having a discussion about their meaning allowed us to better understand what he was trying to say. However, we soon learned there was great latitude in the interpretation of Shakespeare's prose.

The first monologue we studied was from *The Merchant of Venice*, a comedy that had nothing to do with the military—or so we thought. In a scene, a character named Shylock says, "If you prick us do we not bleed? If you tickle us do we not laugh? If you poison us do we not die? And, if you wrong us, shall we not revenge? If we are like you in the rest, we will resemble you in that." During our discussion, one of our members became noticeably quiet. Normally he was very engaging, so his silence brought our attention to him. His head hung low, eyes fixed to the floor, and he fidgeted with his hands.

After a few moments, he told us a story. He had been a lieutenant in Vietnam on patrol when his platoon was attacked from an enemy bunker and a number of his comrades were killed. He spoke softly about how he maneuvered and destroyed the fortified position with grenades. With the firing stopped, he pulled the dead soldier out from the machine gun nest and discovered he was an old Vietnamese man in ragged clothes and sandals. He explained to us that he had held a great hatred for the old man for nearly 40 years because of the lives of his soldiers he had taken. Then he dropped his head again, wiped tears from his face, and said, "Listening to these lines made me realize that the old man was just protecting his family and his land. He was no different than me. I was protecting my men and doing what my country asked of me. I've hated that old man for so long. I forgive him now. I hope his family has forgiven me."

There were similar revelations with the other monologues we studied. Hamlet's famous soliloquy that begins "To be or not to be" generated discussions about soldier suicide; Henry V's St. Crispin's Day speech produced a better understanding of why the bond of the military's brother and sisterhood is so strong and enduring; we also explored whether the character in *Richard III* was not deformed physically but suffered spiritual disfigurement caused by PTSD from the years he spent at war. The dialogue and interactions resulted in such an extraordinary healing process that Dan Minton, a Vietnam War veteran and inaugural member of the group, called it "the best therapy I've had since returning from the war in 1968."

As we became more acquainted with Shakespeare and his writing, the next step was to determine how we would perform his work in front of a live audience. We didn't have the time to prepare and rehearse for a full production with only two hours a week. So, we decided that the performances would be in the choral tradition where each of the members of the group would recite short lines of a monologue. Our movements would be choreographed with the verses that created the effect as if we were speaking in one voice. This appealed to us because it reflected the nature of the military, where battles and wars are won by teams, not individuals.

We would conclude our sessions with a closing question that was always the same. Amy would ask us simply, "What did we learn today?" and an even more in-depth conversation would follow that normally took us well past our allotted two hours. However, our sessions did not always end there. After a time, our group would depart the Vet Center and go to a local bar to continue our discussions, not only about Shakespeare, but about each other and our lives. Within our group babies were born, parents fell ill and died, marriages and relationships ended, anniversaries and birthdays were celebrated, and we shared those times together—just like we did when we were in the military with our brothers and sisters in arms.

On July 17, 2016, our team took the Kentucky Shakespeare stage in Louisville's Central Park and performed scenes from *The Merchant of Venice*, *Hamlet*, and *Henry V* in front of an audience of several hundred spectators. I had been out of the Army for nearly two years by this time, but in the best of ways, it was like I had never left with my new brothers and sisters by my side. I had found my new tribe with Marcus Murray, Cassie Boblett, Rob Givens, Dan Minton, Brian Easley, Patrick Alexander, Dorris Arnold, Jim Meyers, Debbie Sawyer, Aura Ulm, Yvonne Saenz, Al Snyder, Matthew Bone, Steve Montgomery, Steve Gardiner, Darryl Stewart, Stacey Hopson, Alex Banovz, Amy Attaway, and Kyle Ware.

In Act III, Scene I of the play *Henry V*, King Harry encourages his soldiers to continue the fight at Harfleur. He tells them "Once more unto the breach, dear friends, once more." For my brothers and sisters who are transitioning out of the military, that is what we must do. However, it's not on the field of battle that we have to muster the courage to press on. It is out here, in the world, that we must prevail. When we leave the military that is our task and purpose. It's our mission.

Find your passion, something that gives you a sense of purpose bigger than yourself. Discover a way to serve your community, whether it's as a volunteer at middle school or as a first responder. If you are in therapy, stay there as long as it takes to be healthy. Most importantly, know there is always one of us—your brothers and sisters—nearby who will always have your back.

We few, we happy few . . .

ABOUT THE AUTHOR

Colonel Fred Johnson (USA, Retired) has told the story of his journey through *Five Wars* on the stage of the Louisville Moth GrandSlam, with the Louisville Storytellers Project, on the Moth Radio Hour Podcast, and in Reader's Digest.

Since retiring from the Army, Fred has continued his service, now in Louisville, Kentucky, as the Development Officer for Kentucky Educational Television. Fred is also the co-founder of Shakespeare with Veterans, a program dedicated to helping veterans deal with the challenges of transitioning from military service and overcoming PTSD, combat trauma, and moral injury.

Colonel Johnson is from Centralia, Illinois, graduated from Wofford College in Spartanburg, South Carolina, and has two Masters Degrees. While at the US Army War College, Fred was recognized as the top public speaker and received an award for writing the best personal-experience monograph. He is married to Dr. Laura Johnson and they have one daughter, Madelyn Johnson. *Five Wars* is his first book.

You can reach Fred for speaking engagements and author appearances. Contact him here:

EMAIL fredwjohnsonjr74@gmail.com
PHONE (803) 741-4540
WEBSITE fivewars.com

CPSIA information can be obtained
at www.ICGtesting.com
Printed in the USA
FSHW012053080621
82216FS